WASHINGTON'S GENERALS

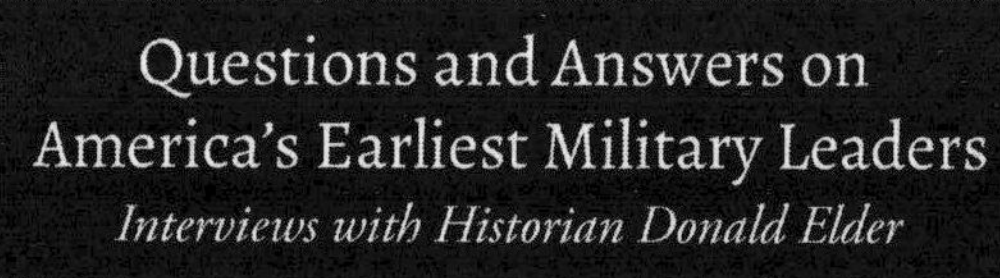

Questions and Answers on America's Earliest Military Leaders

Interviews with Historian Donald Elder

DONALD ELDER AND
MICHAEL F. SHAUGHNESSY

STONE TOWER PRESS

Washington's Generals:
Questions and Answers on America's Earliest Military Leaders

Stone Tower Press
7 Ellen Rd.
Middletown, RI 02842
stonetowerpress.com

Paperback ISBN: 979-8-9868172-5-5

Formatting and cover design by Amy Cole, JPL Design Solutions

Cover image: "Washington's Farewell to Officers, " H. A. Ogden, 1893
Library of Congress Prints and Photographs Division Washington, D.C. 20540
http://hdl.loc.gov/loc.pnp/pp.print

Printed in the United States of America

Table of Contents

Introduction

Victory was far from certain or assured. If anything, most people would have predicted defeat for the American colonists who took up arms in rebellion against British rule. Yet, the Patriot cause prevailed and independence was gained. Under the leadership of General George Washington and his generals the Continental Army conducted military operations throughout the American colonies. With the assistance from the French on land and sea, the Americans prevailed and a new nation was born.

During the course of the American Revolution, in addition to George Washington, the Continental Congress commissioned seventy-seven leaders as general officers. Four of these leaders (John Cadwalader, Seth Pomeroy, Joseph Reed, and John Whitcomb) declined the commission.

Twenty-four of Washington's generals are presented in the following pages. The format is that of questions and answers. Professor of Education, Dr. Michael Shaughnessy has interviewed his colleague at Eastern New Mexico University, Professor of History, Dr. Donald Elder to gain insight and perspective on a remarkable cadre of military leaders. Readers will learn of the proverbial "good, bad, and ugly" regarding generals who led their troops to victory—and sometimes to defeat. Some are well-known and others are not, yet each one left their mark on American history.

WILLIAM ALEXANDER, LORD STIRLING

Painting by Bass Otis, c. 1858
National Portrait Gallery
Image, public domain

ONE

William Alexander

A NOBLE GENERAL ★ (1726–1783)

1. What do we know about when and where William Alexander, also known as Lord Stirling (although some history books have it as the Earl of Stirling), was born?

William Alexander was born on December 4, 1726, in New York City. His father James had moved to North America from Scotland 11 years earlier. While most immigrants came to North America seeking economic opportunity, James Alexander had actually left Scotland seeking sanctuary because of his involvement in a rebellion. In the Glorious Revolution in 1688, the English had ousted King James II and sent him into exile. His son James Francis Edward Stuart in 1715 had attempted to regain the throne through what history calls the Jacobite Rebellion that fully erupted in 1745. James Alexander supported that cause, but when the English defeated the rebels at the 1715 Battle of Preston he realized that the chance to restore the son of James II to the throne had passed.

Accordingly, he booked passage to New Jersey. Fortunately for him, James Alexander had acquired computational skills through service in the British Navy and he utilized that ability to become a surveyor. Soon, he became the surveyor general of New Jersey. This started James Alexander on a rapid rise to prosperity, aided significantly by his marriage to a well-to-do widow. Thus, William Alexander would grow up in a family of wealth and privilege. Using his family connections, William went into business, and became a successful entrepreneur. Much like his father, William married a woman of means—the sister of the governor of New Jersey. By the start of the American Revolution, William Alexander had achieved a position of great prominence in the New York and New Jersey areas and fully supported the Patriot's cause.

2. Why is he often referred to as Lord Stirling?

A distant relative of Alexander's had received the title of the Earl of Stirling from King Charles I in 1633. This title remained in effect until 1739, when the fifth Earl of Stirling passed away. That noble had no son to pass the title to, so in 1756 William Alexander claimed it. He presented his case, based on his distant relationship to the first Earl of Stirling, to a Scottish jury, and in 1759 the body ruled in his favor. From that point on, Alexander used the title of the Earl of Stirling or Lord Stirling. The British House of Lords later reversed the ruling, due to different laws of succession between Scotland and England, thus negating his status of nobility. However, William Alexander continued using the title. He thus joined other members of nobility from other nations in serving the American cause.

3. How did he get involved in the American Revolution?

When news of the battles at Lexington and Concord reached New Jersey, the legislature immediately decided to support the rebellion. In one of their first actions, the body offered William Alexander an appointment as a colonel of the colony's militia. Alexander quickly accepted, demonstrating that he had an immediate infinity for the Patriot cause. Although we have no definitive proof, we think two factors helped to influence his decision. First, as previously mentioned, Alexander's father had come to North America after participating in a rebellion against the English government.

His father's experience undoubtedly shaped the feelings of Alexander toward the monarchy, and in all likelihood helped guide him to the Patriot cause. And second, his brother-in-law had sided with the Patriots, and this family dynamic probably factored into his decision. Whatever his motivation, Alexander clearly threw his lot in with the rebels, and remained devoted to the cause throughout the Revolutionary War. Interestingly, he also had served in the French and Indian War (1754–1763) but in its aftermath he became opposed to British colonial policies.

4. How long did he actually serve in the Revolutionary War?

As we will see, Alexander found himself in the thick of the fighting that George Washington did in the first three years of the Revolutionary War. From that point on, however, Alexander saw little actual combat. He commanded a raid on Staten Island in January of 1780 that had little effect, and when Washington led the bulk of his army to Virginia in 1781, he left Alexander in front of New York City as a diversion. After the British surrender at Yorktown in October of 1781, Washington returned with the majority of his army to the outskirts of New York City and Alexander once again became his subordinate. For the next year, the British and American armies kept each other at bay in that region. Unfortunately for Alexander, rheumatism and gout began to take a heavy toll on his health. Sadly, he died in January of 1783, a few months before news of the Treaty of Paris ending the war reached North America.

5. Apparently, he was very involved in a number of battles in the New York, New Jersey area. Can you comment a bit about his role in each of the following battles?

- Battle of Long Island (August 1776)
- Battle of Trenton (December 1776)
- Battle of Brandywine (September 1777)

- Battle of Germantown (October 1777)
- Battle of Monmouth (June 1778)

Once he became a colonel of the New Jersey militia, Alexander used his own fortune to equip a regiment of soldiers. Showing great daring, he led his troops on an attack against a British transport ship. His capture of that vessel came to the attention of the Second Continental Congress, and they rewarded Alexander with an appointment as a brigadier general in the Continental Army in the summer of 1776.

After Congress gave Alexander his commission, he joined George Washington's army in its encampment in and around New York City. Given command of a brigade (numbering 3,700 men and consisting of soldiers from Maryland, Delaware, and Pennsylvania), Alexander positioned his force on Washington's right flank on Long Island. On August 27, a large British army attacked the Americans, and through a skillful maneuver they managed to outflank Washington's position. Most American soldiers fled in panic or surrendered, but Alexander chose to hold his ground. He ordered most of his soldiers to retreat, but kept four companies of the 1st Maryland Infantry in position.

As the British approached the Marylanders, Alexander ordered his men to attack the enemy. This stopped the British advance and forced them to regroup. Eventually, the British attempted to move forward, but once again Alexander's men stopped them with an attack. This pattern was repeated four more times, until Alexander realized the British had virtually surrounded him. Ordering his men to retreat, Alexander tried to flee from the British but wound up having to surrender to a detachment of Hessians (soldiers from present-day Germany whose rulers had sent them to America to fight with the British in return for money). Only nine of his men escaped death or capture, but Alexander's courageous stand allowed Washington to get his army to safety.

Alexander did not remain a prisoner for long. Earlier, in the spring of 1776, American forces had launched a raid on Nassau in the Bahamas, and during that action they had captured the royal governor. The Americans offered to return him in exchange for Alexander. This trade took place in the fall of 1776. Impressed with the tenacity and fighting spirit that Alexander had displayed at Long Island, Congress promoted him to major general upon his return to duty. Alexander rejoined the army as it retreated from New York through New Jersey and into Pennsylvania.

During the retreat, Washington's army had shrunk to only 3,000 men, and few gave the Patriot cause much chance of succeeding. Recognizing that he needed to boost morale, Washington decided to cross the Delaware River and attack a brigade of Hessians stationed at Trenton, New Jersey. Accordingly, on December 26, 1776, Washington launched a surprise assault at daybreak. Once again in command of his brigade, Alexander attacked the Hessians from the north and soon succeed in routing the Hessians in front of him. In short order, and with minimal casualties, Washington succeeded in killing or capturing all the Hessians, giving the Americans an impressive victory. Washington's daring initiative of the crossing of the Delware River is majestically portrayed in the famous 1851 oil-on-canvas painting by Emanuel Leutze.

In 1777, the British launched a campaign to capture Philadelphia, at the time the nation's de facto capital. Washington moved his army into a defensive position outside of the city, deploying his forces along Brandywine Creek. Unfortunately for Washington, the British employed the same strategy that they had used on Long Island the year before, and by the end of the day the Americans once again found themselves fleeing the battlefield.

And just like at the Battle of Long Island, Alexander tried to stave off disaster by holding his position long enough for Washington to withdraw the rest of his army. Here again, Alexander succeeded in that task. After their victory September, 1777, the British occupied Philadelphia, but Washington devised a strategy to try to regain the city. Recognizing that the British had moved a portion of their army to the nearby community of Germantown, Washington decided to throw his entire army against that detachment. In the ensuing battle, Alexander's division did not see action. Rather, Washington held them back as his reserve, planning to deploy them to exploit any breakthrough. Because no such opportunity presented itself to Washington, Alexander's troops never went into battle at Germantown.

After the defeat, Washington went into winter quarters at Valley Forge for the winter of 1777–1778. Later, when Washington learned in June of 1778 that the British had decided to abandon Philadelphia, he decided to attack the enemy as they marched back to New York City. After examining the line of the British march, Washington selected the area around Monmouth Court House in New Jersey as the location to focus his assault. Unfortunately for Washington, a subordinate did not execute the initial attack according to plan.

Worse, that section of the American army began a precipitous withdrawal! Alerted to this retreat, Washington brought up the rest of his force, and deployed Alexander's division on his left flank. Throughout the rest of the battle, Alexander and his men thwarted numerous British attacks, and by the end of the day they still held their initial position. That night, June 28, 1778, the British chose to leave the battlefield, giving Washington a strategic victory, one aided in large measure by Alexander's stalwart stand. As it turned out, the Battle of Monmouth Court House was the last major engagement for General Alexander.

6. What was William Alexander's major contribution to the American cause?

Actually, he gave invaluable aid to the American cause through an action far removed from the battlefield. After Washington had failed in his efforts to defend Philadelphia in 1777, a number of high-ranking American officers began to discuss a plan to take command of the army away from him. Led by Brigadier General Thomas Conway, this movement comes down through history known as the Conway Cabal. The content and details of the letter that circulated among the dissatisfied officers came Alexander's awareness and he immediately forwarded it to Washington. Armed with this information, Washington moved quickly and effectively to quash the insurrection, and no one ever challenged his authority again. Alexander's decisive action regarding the Conway Cabal thus might have had an even greater impact on the outcome of the war than did his courageous stand at the Battle of Long Island. Though he died before the war ended, he was one of Washington's most loyal generals.

BENEDICT ARNOLD

Engraving by H. B. Hall, 1865
Library of Congress
Image, public domain

TWO

Benedict Arnold

TRAITOR OF THE AMERICAN REVOLUTION ★ (1741–1801)

---

1. Almost all Americans know Benedict Arnold as a traitor to America. Where was he born, and what do we know about his early life that may give us some clues as to his treachery?

Benedict Arnold was born in Norwich, Connecticut on January 14, 1741. For the first few years of his life, he enjoyed a privileged existence, as his father ran a successful business in Norwich. His family's wealth allowed Benedict Arnold to attend a private school in Canterbury, Connecticut, and the family planned for him to become a student at Yale College after he finished his primary and secondary schooling.

Unfortunately, tragedy soon struck the family, as three of his siblings died in a five-year span in the early 1750s. These deaths had a deleterious effect on his father, causing him to drink to excess. His father's consumption of alcohol and health troubles resulted in his neglecting the family business, and as a result Benedict had to leave school. Although his father became a social pariah, Arnold's mother proved able to arrange for Benedict to become an apprentice to an apothecary for eight years during his teenage years. Soon after he assumed that role, the French and Indian War widened into the conflict known in Europe as the Seven Years' War. Connecticut began to recruit soldiers to fight the French and their Native American allies, and in 1757 Benedict Arnold answered the call.

Soon, he and his comrades received orders to proceed to New York to relieve the siege of British-held Fort William Henry at the southern end of Lake George. Once on the march, however, they received the news that the garrison had surrendered, and that Native Americans had killed many of the defenders after the French had guaranteed them safe passage after they laid down their arms. When they learned of this turn of events, Arnold and his fellow volunteers turned around and went home.

During the American Revolution, Arnold would show great bravery on the field of battle, but his willingness to give up without a fight during the French and Indian War may have presaged his abandonment of the American cause in 1780.

2. What did his early military endeavors involve?

Sensing that hostilities with Great Britain might begin at any moment, in March of 1775 the Connecticut community of New Haven recruited a company of militia, naming Benedict Arnold the captain. In the next few weeks, Arnold threw his energies into training his unit. When he learned of the Battles of Lexington and Concord, Arnold decided to take his militia unit to join the Patriot force besieging the British in Boston. As he journeyed northward, Arnold had a meeting with a member of the Connecticut legislature in which the two talked about the lack of cannon among the forces gathered outside Boston.

During their discussion, they discussed the fact that Fort Ticonderoga in northern New York had a significant number of cannon, and they agreed that Connecticut should send a force there to capture those weapons. The legislator left to arrange for volunteers to accomplish that task and Arnold resumed his march to Boston. When he reached his destination, Arnold talked to Massachusetts officials about sending another force to Fort Ticonderoga to reinforce the detachment already headed there. Impressed with his proposal, Massachusetts gave him a commission as a colonel of its militia, and gave him permission to recruit volunteers to participate in his campaign.

When he reached western Massachusetts, sources informed him that an individual named Ethan Allen had already reached Fort Ticonderoga, planning a mission similar to Arnold's. Quickly making his way to upstate New York to reinforce Allen, Arnold found Allen poised to launch an attack the following day. At first, Arnold tried to assume command of the operation, based on his instructions from the authorities in Massachusetts. But Allen's men, known as the Green Mountain Boys, refused to follow Arnold's orders. Arnold countered with a request for a joint command and it was accepted.

Accordingly, Allen and Arnold both led an attack on Fort Ticonderoga on the night of May 10, 1775. Catching the British completely by surprise, the Americans captured Fort Ticonderoga without suffering a single casualty. This victory gave the Patriots cannon that George Washington would use to force the British to evacuate Boston in March of 1776.

3. In his early Revolutionary War battles, what were Arnold's greatest accomplishments?

After helping capture Fort Ticonderoga, Arnold remained there to administer the installation. Soon, however, a force sent to garrison the fort arrived, led by a colonel who bore orders giving him command of Fort Ticonderoga. Angered by this, Arnold resigned his commission and returned to Connecticut. Unable to remain inactive, Arnold went to Washington's camp to propose a campaign. Recognizing the strategic value of Canada, in the summer of 1775 the Continental Congress had authorized an expedition to secure that region.

In his meeting with Washington, Arnold suggested that he could provide a second prong for the Canadian campaign by moving overland through present-day Maine. Impressed by the proposal, Washington made Arnold a colonel in the Continental Army and gave him a force of 1,100 soldiers. After a harrowing march similar to that of George Rogers Clark in the American hinterland, Arnold reached Canada in November of 1775 after losing half his men. Linking up with the main American force under the command of General Richard Montgomery, Arnold and his men participated in an assault on Quebec on December 31, 1775. Leading a charge, Arnold received a serious leg wound.

Refusing evacuation, Arnold remained on the battlefield to direct the assault until it became obvious that the attack had failed. He then allowed his men to carry him to safety. Although the American attack had failed and General Montgomery had died in the process, Arnold kept his depleted force in the vicinity of Quebec, hoping to resume his efforts to capture the city. The arrival of British reinforcements in Canada made this impossible. Indeed, Arnold found it necessary to abandon Canada altogether. As he retreated south, a large British force pursued him. Knowing that the British would use Lake Champlain to transport their troops, Arnold ordered the construction of a small fleet to stymie the British. Responding in kind, the British built their own flotilla, and sallied forth to meet Arnold's fleet.

At the Battle of Valcour Island in October of 1776, the British decisively defeated Arnold, but the naval activities had so delayed the British advance that their troops had to retreat to Canada before winter set in. Promoted to brigadier general, Arnold received orders to report for duty in his home state of Connecticut. There, he received another leg wound while leading his men into action at the Battle of Ridgefield in April of 1777. Fortunately for Arnold, this wound healed quickly, and did not prevent him from accepting an assignment to the Northern Army gathering to oppose the combined British-Hessian force that had invaded New York in the summer of 1777.

Once again, Arnold provided inspiring leadership in battle, especially in spearheading a charge at Bemis Heights on October 7, but once again his bravery led to a leg wound. Of his three leg wounds, this one proved the most serious. As a result of it, Arnold lost two inches of his left leg, and he would walk with a pronounced limp for the rest of his life.

4. Apparently, he had some disagreements with Washington. What were some of the issues?

Enlisted men serving under Arnold gave him their adulation, but he had conflicted relationships with many of his fellow officers. In addition, a number of powerful civilian leaders found Arnold vainglorious. Some of his personal interactions proved so toxic that he had to face courts of inquiry. Throughout all of the turmoil, George Washington remained steadfast in his defense of Arnold. Indeed, when Arnold resigned after Congress passed him over for promotion to major general, Washington personally intervened with the legislative body to rectify the situation. But events that took place in 1778 placed Washington into an awkward situation regarding Arnold.

Because of the wound that he had suffered at Bemis Heights, Arnold could not command troops in the field, and so after the British evacuated Philadelphia in 1778 Washington gave him command of the reoccupation process. As the military governor, Arnold became involved in some highly questionable financial ventures involving confiscated Loyalist property, and in February 1779 he learned that he would face a court of inquiry regarding these transactions. When the military tribunal found him guilty on two charges, Washington felt compelled to give him a rebuke in writing. Carefully phrased to accentuate his meritorious service and couched in mild terms, the reprimand nonetheless stung Arnold to the quick, and forever altered the relationship between the two.

5. Major John André and, of course, Peggy Shippen were key figures in his treason. What part did each play?

During the British occupation of Philadelphia, British Major John Andre had become acquainted with Peggy Shippen, the beautiful daughter of a merchant with Loyalist sympathies. Questions about their level of intimacy exist, but it seems clear that the two held great affection for each other, as upon her death it turned out that she had kept a lock of André's hair. After the British evacuated the city, Arnold courted and wedded Peggy Shippen, but she kept in touch with André through a middleman named Joseph Stansbury. This connection proved beneficial to Arnold when he made his decision in the spring of 1779 to betray his country, as he conveyed his intentions to the British through that go-between.

6. West Point is a place that really needs no introduction, but what role did it play in this entire nefarious affair?

After his rebuke from Washington, Arnold resigned his position as the military governor of Philadelphia. Hoping to once again make use of Arnold's talents in battle, Washington offered him the opportunity to command troops in the field. But Arnold declined, asking instead for the command of the American fort at West Point, north of New York City on the Hudson River. Established to prevent easy passage of a British force either up or down the river, many considered it the most important American fort. Puzzled by Arnold's request for a stationary assignment, Washington nonetheless granted it in August of 1780.

A month later, Washington received a communique from Arnold, inviting him to inspect the fort's defenses. Washington agreed, and arranged a visit with Arnold. Upon his arrival at West Point, Washington found the fort in a dilapidated condition and Arnold missing. Soon, Washington learned that Arnold had deliberately allowed the defenses to deteriorate, as he had promised the British that he would do that to make its capture easy for them. In addition, he had arranged for Washington to visit at the same time that the British would attack West Point, thus giving them the most important fort and leader in one fell swoop. John André had served as the intermediary between Arnold and the British military, but three American militiamen had captured him with the plans on his person.

They then took André to Arnold at West Point. Recognizing that the plan would no longer work and that André could tell the Americans of his treachery, Arnold ordered him imprisoned and immediately left West Point. After alerting his wife to the fact that he had to leave, he had himself rowed to a British warship lying in wait on the Hudson. By the time Washington got to the bottom of the situation, Arnold was far beyond his reach.

7. Apparently he escaped to England with Peggy Shippen. What do we know of his later years?

In return for betraying his country, Arnold received an appointment as a brigadier general in the British Army and a large sum of money. In 1781, he led British troops in action against his former countrymen in Virginia and Connecticut. After the surrender of Cornwallis's army at Yorktown, it became obvious that the British would not be able to militarily subdue the Americans. For that reason, Arnold and his wife (whom Washington had magnanimously allowed to join Arnold after his treachery became known) sailed to Great Britain. They would live there for the rest of their lives.

8. Obviously, his early years were filled with some positive accomplishments, but his betrayal will live forever, as they say, in infamy. But what have we learned from the life and betrayal of Benedict Arnold? And what ever happened to John André?

After meeting with Arnold at West Point to arrange the capture of the fort and the visiting George Washington, Arnold had convinced John Andre to make his way back to the British lines in civilian clothing. This proved fatal for Andre when the militiamen captured him, because it meant that the Americans considered him a spy instead of a captured enemy officer. The military tribunal that tried André sentenced him to death by hanging. Learning of this, the British military pleaded with Washington to spare the popular major's life. Washington replied that he would gladly do so—if the British returned Benedict Arnold.

This the British refused to do, and Washington then ordered André's execution by hanging. Many British officers had a distaste for Benedict Arnold for betraying his comrades, and the fact that he had lived while a British officer died in his stead only worsened things for Arnold. After the war ended, Arnold never received an important military assignment, and all of his post-war business ventures failed. Few therefore seemed distraught when he died in 1801.

9. In sum, what can we say about this person?

It is extremely hard for Americans to come to terms with the treason of Benedict Arnold. How could a man who fought so valiantly for a cause suddenly betray it? Was his treachery due to the fact that he never felt sufficiently recognized for his contributions, or from his anger over the legal challenges that he faced throughout his military career? Did Peggy Shippen cause him to abandon his loyalty to the United States, or did she merely facilitate his betrayal? Interestingly, he was the only man to serve as a general on both sides of the war. Perhaps the only thing certain about Benedict Arnold is that if he had died of his wound at Bemis Heights, he might have gone down in history as the greatest of all of Washington's generals rather than one of America's greatest traitors.

GEORGE ROGERS CLARK

Painting by James Barton Longacre, 1825
National Park Service
Image, public domain

THREE

George Rogers Clark

A SELF-DESTRUCTIVE HERO ★ (1752–1818)

---

1. George Rogers Clark was a very famous general who actually had a stamp created to commemorate the 150th anniversary of his 1779 victory at Fort Vincennes. Where and when was he born?

George Rogers Clark was born on a plantation in Albemarle County, Virginia on November 19, 1752. His family did not remain there for very long after his birth, however. When hostilities between Virginia and the French and their Native American allies began in 1754, Clark's father moved his family further east to Caroline County to escape the danger of a raid on their property. George Rogers Clark then grew to manhood in that more eastern part of the colony.

2. On an interesting historical note, where did he go to school, and who did he go to school with as a classmate?

Since Caroline County did not have a public school, Clark's father arranged for him to live with his grandfather in King and Queen County, Virginia, where one did exist. A Scotsman named Donald Robertson had moved to that county in 1758, and opened a school. George Rogers Clark attended that school, and had a classmate who would become famous after the American Revolution—the future US President James Madison.

3. What do we know of his early forays into battle as part of the militia?

In addition to the education he received at Donald Robertson's school, George Rogers Clark also received instruction in the art of surveying from his grandfather. He began to put that skill to use at the age of 19 in the area now known as West Virginia. A year later, he journeyed further west into present-day Kentucky to survey land there. Many colonists moved to this region during this period and the migration angered the Native Americans who hunted there. This animus led to the Native Americans killing a number of settlers. Retaliatory attacks by colonists soon followed, and in May 1774, Lord Dunmore, the governor of Virginia, called upon the colonial legislature to declare war on the tribes responsible for the initial attacks.

These events happened as George Rogers Clark stood poised to lead a party of settlers into Kentucky. At the start of the conflict, soon named Lord Dunmore's War, Clark entered the Virginia militia as a captain. He saw no actual combat before the war ended in October 1774, but learned valuable lessons about military affairs during his service.

4. Kaskaskia and July 4 are almost synonymous with his name. Why?

When Lord Dunmore's War ended, George Rogers Clark followed through on his plan to help settlers reach Kentucky. Once he arrived there, however, he found that a jurisdictional dispute had developed regarding who had the authority to govern the Kentucky area. By the spring of 1776, a number of the residents decided that they wanted to become a part of Virginia. This group convinced Clark and a man named John Gabriel Jones to go to Williamsburg to ask the Virginia legislature to make Kentucky a county. When they arrived, they accomplished their goal, Kentucky County was established (and later admitted as a state in 1792 after being split into nine counties), but they also received a new responsibility.

By then, hostilities between Great Britain and the thirteen colonies had commenced, and Virginia had committed most of its troops to that fight. Patrick Henry, the governor of Virginia, told Clark and Jones that meant that he could not supply troops to help defend the frontier if trouble with Native Americans developed. Instead, the residents of Kentucky County would have to defend themselves. To help in that regard, Henry gave Clark a commission as a major in the militia, and gave him a supply of gunpowder to take back with him. Arriving back in Kentucky, Clark soon had good reason to appreciate that gift, as Native Americans allied with the British began attacking settlers throughout the region. Clark spent a year fending off the Native Americans.

As he did, he realized that to end the incessant raids he would need to capture three British forts north of the Ohio River where the Native Americans received muskets and ammunition for their incursions. During a lull in the fighting in December of 1777, Clark went back to Williamsburg to ask Governor Henry for permission to launch a campaign to accomplish his goal. Obtaining Henry's consent and a promotion to lieutenant colonel, Clark then recruited a force of approximately 200 men to accompany him.

In July 1778, Clark took his men into present-day Indiana, with the goal of capturing a British fort at Kaskaskia. Arriving on July 4th, Clark convinced the British that they could not possibly hold the fort against a determined American assault. Recognizing the hopelessness of their situation, the British therefore surrendered that night without offering any resistance. Just as General Ulysses S. Grant would do during the Civil War with his capture of Vicksburg on July 4, 1863, George Rogers Clark did with Fort Kaskaskia and thus gave the nation a fitting birthday present by taking the fort on Independence Day in 1778.

5. Some historians credit Clark with greatly expanding the 13 colonies. What endeavors of his led to this?

A day after Clark captured Fort Kaskaskia, a portion of his expeditionary force compelled Cahokia, a British fort located in present-day Illinois, to surrender. With the capitulation of Fort Vincennes in Indiana later that summer, Clark had succeeded in accomplishing the goals of his campaign. His success proved short-lived, however, as the British and their Native American allies managed to retake Fort Vincennes in December of 1778. Learning of this, Clark decided on a risky course of action.

Realizing that the British and Native Americans would undoubtedly remain at Fort Vincennes until the spring, Clark marched his men through extremely inclement weather from Kaskaskia to Vincennes in order to make a surprise attack before they could move elsewhere. In spite of the perilous conditions, Clark and his men reached Vincennes in February of 1779 and laid siege to the outpost. At first, the British commander refused to surrender, hoping that Native American reinforcements could arrive in time to break the siege. Clark thwarted this by defeating the relief force, capturing two Native American leaders in the process.

Taking the two leaders to a clearing near the fort, Clark had them publicly executed. That night, many Native Americans inside Fort Vincennes deserted, and the next day the British commander surrendered. At the end of the American Revolution, the Americans still held these three forts. That allowed the Americans negotiating the treaty to end the war to claim all the territory from the Atlantic Ocean to the Mississippi River by right of possession. Thus, Clark and his small force secured a huge expanse of territory for the new nation that it otherwise could never have asked for in the treaty.

6. It was Thomas Jefferson who appointed him brigadier general. What did he do to deserve such an honor?

Although the British posed no serious threat to Kentucky after the capture of Vincennes, Kaskaskia, and Cahokia, Native Americans continued to attack settlers there. Given the task of defending Kentucky, Clark decided to strike the Native Americans north of the Ohio River before they could reach their target destinations. Meeting a sizable force of Native Americans near present-day Springfield, Ohio, Clark won a decisive victory in August of 1780. Although his triumph did not end violence on the frontier, it did give Kentucky temporary respite from the dangers it had hitherto faced. In recognition of his contributions to the pacification of the frontier, in 1781 Thomas Jefferson, by then the governor of Virginia, promoted George Rogers Clark to the rank of brigadier general. Until the American Revolution ended in 1783, with that rank Clark continued to lead campaigns against Native American tribes on the frontier.

7. Apparently he received a number of nicknames during the American Revolution. Some of them were: Conqueror of the Old Northwest, Father of Louisville, Hannibal of the West, and Washington of the West. How did all these names come about and what did he do to deserve such recognition?

As we have seen, Clark won a series of victories against British and Native American forces in what was known as the Northwest Territory during the American Revolution. These achievements earned him the nicknames associated with military endeavors. In passing, it could be noted that comparing Clark with Hannibal does not do the American justice, as the Carthaginian general ultimately suffered defeat and death in the Second Punic War. Clark had no such major setback during the American Revolution. As it turns out, the nickname "Father of Louisville" also resulted from the American Revolution.

This came about because when Clark began his campaign in 1778 to subdue the British forts, 80 settlers traveled with him as far as the Falls of the Ohio River. Recognizing the advantages offered by the topography there, Clark suggested that the settlers start a community at that location. Because France had become the ally of the United States in 1778, the settlers named their new town after King Louis XVI. Clark thus deserves the title of "Father of Louisville" just as much as he does the nickname "Conqueror of the Old Northwest."

8. Sadly, like many political and military leaders of the American Revolution, his later years were not kind to him. Can you summarize his later years?

Although hostilities with the British ended in 1783, conflict with Native American tribes lingered on for many years. In 1786, Clark (who had chosen to live on the frontier after the American Revolution) led a force into present-day Indiana to subdue the tribes that had refused to make peace with the United States. But unlike his campaigns against the British and Native Americans during the American Revolution, Clark proved unable to win a decisive victory. A number of individuals suggested that Clark's failure stemmed from his growing predilection for alcohol.

Angered by these rumors, Clark asked for a chance to defend his honor through a hearing. Virginia's governor refused his request, however, seemingly giving credence to the accusations of drunkenness. To make things worse for him, Clark soon found himself facing substantial debt. To finance his campaigns during the American Revolution, Clark had signed personal notes, confident that Virginia and/or Congress would reimburse him for his expenses.

But because Clark kept inadequate records, both Virginia and Congress refused most of his requests for recompense. Virginia did give him 150,000 acres of land, but Clark did not have the financial resources to improve his property. This made his holding essentially worthless. To pay off his creditors, Clark eventually had to cede much of his land to them. Felled by a stroke in 1809, Clark soon thereafter lost his leg to amputation due to a severe burn he suffered after falling into his fireplace. He spent his last few years living with his sister and brother-in-law on a farm near Louisville. He died there in 1818.

9. For historians and those who love American history, when we see the name George Rogers Clark we immediately think of the Lewis and Clark expeditions that we studied as students. Were these two related?

George Rogers Clark came from a large family, one that certainly helped to shape the history of the United States. First, four of Clark's brothers also fought for the Patriot cause during the American Revolution. All of them served as officers. Born in 1770, Clark's youngest brother was the only one that did not serve during the American Revolution, but he did receive a commission in the Kentucky militia in 1789. Three years later, he enlisted in the US Army, and received an appointment as a lieutenant. After earning praise for his performance at the Battle of Fallen Timbers in 1794, he resigned his commission two years later. Up to that point, the youngest Clark had done nothing to surpass the fame earned by his older brother George, but all that changed in 1803 when his friend Merewether Lewis asked him to help command an expedition to explore the Louisiana Territory.

As many readers will have by this point in time no doubt guessed, the youngest Clark son was William Clark (1770–1838), who because of the Lewis and Clark Expedition to the Pacific Northwest (1804–1806) would become one of the most famous Americans of the 19th century and beyond. William Clark's success, coupled with the troubles that had befallen his older brother, led to George Rogers Clark falling into obscurity. Since his death in 1818, George Rogers Clark's name has gradually reemerged in American history, especially in the area of the country that he helped wrest from British control during the

American Revolution. This resurrection of his stature is completely warranted, as the United States would have occupied a much smaller space in 1783 than it did without his remarkable campaign in 1778–1779. Hopefully, his name will continue to be reestablished in the panoply of Revolutionary War heroes.

GEORGE CLINTON

Painting by Ezra Ames, 1814
Image, public domain

FOUR

George Clinton

A LEADER WEARING MANY HATS ★ (1739–1812)

1. For many historians, the name Clinton goes back many decades, if not centuries. Regarding George Clinton, when and where was he born? And what do we know about his parents?

George Clinton was born on July 26, 1739, in Little Britain, New York. At the time of Clinton's birth, Little Britain lay in Ulster County. His parents, Charles and Elizabeth Clinton, had emigrated from County Longford in Ireland in 1729, and soon became prosperous residents of Little Britain. Indeed, the elder Clinton became a surveyor and land speculator, and won election to the New York General Assembly. During the French and Indian War (1754–1763), Clinton's father received a commission as a colonel in the New York militia. Following the example of his father, George Clinton chose to take part in the French and Indian War. After initially serving on a privateer, he offered his services to the New York militia. He soon rose to the rank of lieutenant, and in 1758 he and his father both participated in a campaign that resulted in the capture of Fort Frontenac on Lake Ontario. While George Clinton would eventually become much better known than his father, clearly Charles Clinton gave his son an enviable record to try to better.

2. Clinton apparently was a governor and a lieutenant governor and general all at the same time. Talk about working overtime, that's impressive! How did this come about?

When hostilities broke out between Great Britain and its North American colonies in April of 1775, New York appointed George Clinton (at the time serving in the state assembly) brigadier general of militia. Colonial officials then gave the responsibility of protecting New York City from attack through the Highlands region north of that community. In March 1777, he received the rank of brigadier general in the Continental Army. Three months later, New York held its first election for the positions of governor and lieutenant governor. Interestingly, the state's electorate chose him for both offices. Upon learning the outcome of the election, Clinton resigned as lieutenant governor, and then accepted the position as governor. Thus, for a brief moment in time George Clinton did simultaneously hold the title of lieutenant governor, governor, and general.

3. He was apparently very helpful to Washington at Valley Forge. What were his specific contributions?

Most Americans, when they hear the words "Valley Forge," immediately picture an American army suffering in miserable conditions during the Revolutionary War. In large measure, this image comes from a letter that George Washington wrote in February of 1778 describing the hardships that his men faced there. As it turns out, George Washington addressed that February 16, 1778 letter to the governor of New York—George Clinton. Even though, as Washington himself noted, New York lay far away from Valley Forge, Clinton did his utmost to send badly needed aid to Washington.

4. He was known for two forts and for his efforts to contain the British from navigating the Hudson River. What were his contributions there?

In 1777, a combined British and Hessian force under the command of General John Burgoyne invaded the American colonies from Canada. Burgoyne envisioned transporting his force by water down

Lake Champlain, and then advancing the troops by land along the Hudson River. If he had succeeded in his plan to eventually reach New York City, he would have effectively cut New England off from the other states. As Burgoyne moved south, he faced increasingly resolute resistance from American forces, and by mid-September found himself bogged down near Saratoga, New York. Hoping to continue his advance before winter set in, Burgoyne called upon the British military units stationed in New York City to create a diversion to help him break out. To that end, a British unit under the command of Sir Henry Clinton began a maneuver up the Hudson River towards Burgoyne's stranded force. Standing in Clinton's way stood two American forts: Clinton and Montgomery. The American defenses also included an iron chain strung across the river, and four American vessels used to slow or block British ship traffic on the Hudson River.

Unfortunately, a large portion of the force assigned to defend the forts fell for a British diversion, and moved into the Highlands to meet the perceived threat. This left only 600 men, under the command of George Clinton and his brother James, to defend the forts. When Sir Henry Clinton attacked the forts on October 6, 1777 with 3,000 men, the American force guarding the installations proved too small to defend them. Sir Henry Clinton took both forts by nightfall, and killed or captured 350 of the defenders at a cost of only 190 total casualties. But although he lost the forts, George Clinton helped slow the British relief force sufficiently to the point that Sir Henry Clinton could not reach Burgoyne's isolated command before it capitulated on October 17. The surrender of Burgoyne's army proved decisive, as news of it helped convince France to become an ally of the United States.

5. He was well known for his hatred of Tories. To refresh our memories, who were the "Tories" and what did he do to harass them?

No one can say with absolute certainty what percentage of the population of the 13 rebelling states actively supported the cause of independence. In a similar fashion, no one knows how many colonists remained committed to the crown. Whatever their numbers, individuals who fell into the latter category come down through history referred to as either "Loyalists" or "Tories." Americans today can easily understand the former term, but undoubtedly find the latter term puzzling. It actually comes from British politics. In the 17th century, two schools of thought had coalesced regarding the use of power by British monarchs. Individuals favoring the aggrandizement of power by the crown became known as Tories. (The others were known as Whigs.)

It therefore seemed logical to Americans favoring independence to brand their compatriots still loyal to King George III as Tories. While every state had some Tories, New York probably had the most, and many of them actively fought against the cause of independence. This partisan warfare hardened feelings on both sides in the Empire State, and makes Clinton's hatred of Tories much more understandable.

6. Sadly, he missed signing the Declaration of Independence. What was he doing at the time?

After serving in the New York assembly in 1775, Clinton's fellow citizens chose him to serve in the Continental Congress the following year. Some individuals sent to Congress in 1776 threw themselves enthusiastically into their work, but Clinton split his time attending to his military and political duties. As fate would have it, Clinton found himself in New York at the time that Congress drafted and then approved the Declaration of Independence. Had he remained in Philadelphia that summer, his name would thus be on the famous document.

7. One of the last accomplishments I have to mention is what occurred at Dobbs Ferry, New York, home of the Mercy College Flyers-and a great college. Apparently, it was at Dobbs Ferry that Clinton and Washington finally were able to get all of the British troops out of the United States. What do we know about those negotiations?

After General Charles Cornwallis surrendered the force under his command at Yorktown, Virginia in October of 1781, the British government had recognized the futility of continuing its attempt to suppress the rebellion in North America. As a consequence, peace negotiations had then begun, and finally on September 3, 1783, the Treaty of Paris ended the hostilities. One provision mandated that the British would evacuate their forces from the newly independent United States. As it turns out, the British government had recognized that the treaty would undoubtedly call for this, and in August had ordered Sir Guy Carleton, the commander of British forces in North America, to facilitate that process. Carleton then contacted George Washington to arrange for an orderly transition of power. The two chose Dobbs Ferry, New York as the place that they would meet to discuss procedures. Because he served as both a general and the governor of New York, George Clinton received an invitation to the meeting from Washington.

At Dobbs Ferry, Carleton agreed to remove his troops from New York City as soon as possible, but then he surprised the Americans by refusing to abide by one provision of the treaty. During the American Revolution, thousands of slaves had left their masters and mistresses, seeking sanctuary from the British. They did so because the British had promised slaves their freedom if they left their owners. In light of the promise made by his government, Carleton informed Washington and Clinton that he could not in good conscience agree to return the runaway slaves (a number of whom had subsequently enlisted in the British Army).

This displeased Washington, but he grudgingly accepted Carleton's promise that he would forward a list of the fugitive slaves to London so that American owners could seek compensation for their losses. Historians believe that no slave owner ever received compensation from the British government. Washington could have chosen to dispute Carleton on his failure to honor the "letter of the law," but although he personally had suffered the loss of slaves himself, he felt that achieving the goal of a British evacuation outweighed the sleight administered by his rival. Washington, Clinton, and every other American at the meeting at Dobbs Ferry thus deserve high marks for seeing the big picture.

8. What have I forgotten to ask about his contributions to the American Revolution and his assistance to George Washington?

In our series of articles on Washington's generals, we have extolled the virtues of high ranking officers who gave him military assistance during our nation's fight for independence. We could devote a series of equal length to the political leaders who helped sustain the fight by mobilizing the civilian population to support the cause. George Clinton stands in rarified company as an individual that would merit an article in both series. Add in the fact that he is one of only two Americans to ever serve as vice-president for two different presidents (Thomas Jefferson and James Madison), and George Clinton clearly emerges as one of the most multi-faceted individuals in our history!

HORATIO GATES

By Charles Wilson Peale, 1782
Independence Hall, Philadelphia (INDE 14053)
Image, public domain

FIVE

Horatio Gates

HERO AND COWARD ★ (1727–1806)

1. What do we know about the early life General Horatio Gates?

Like many American generals who supported the Patriot cause during the American Revolution, Horatio Gates came here from another land. But while most of these individuals arrived in North America during the war, Gates came here from his native Great Britain long before the outbreak of hostilities. Born on July 26, 1727 in Maldon, England, Horatio Lloyd Gates grew up in a fairly privileged position due to the connections that his parents had made through employment with members of the English nobility.

Because of this background and with financial assistance from his parents and some patronage, Gates purchased and received a commission as an ensign in the British Army at the age of 18 in 1745. Assigned to an infantry regiment (20th Regiment of Foot), Gates saw military action in present-day Germany as part of the War of the Austrian Succession (1740–1748). At the conclusion of that conflict, the British government transferred Gates and his regiment to North America. After serving there for a year, Gates received a promotion to the rank of captain and a transfer to a different infantry regiment. In 1750, he participated in a battle in Canada against Native Americans loyal to France in what was known as the Micmac War (1749–1755) and also in the French and Indian War.

In 1755 he accompanied British General Edward Braddock on his ill-fated expedition into the Ohio Territory against Fort Duquesne, and received a serious, but non-life threating, wound during that campaign at the Battle of the Monongahela (July 9, 1755). Ironically, Braddock's expedition also included George Washington, who Gates would serve under two decades later. Braddock's foray into the Ohio Territory provoked a war between the British and the French, a conflict called the French and Indian War (1754–1763) in the United States and known in Europe as part of the Seven Years' War. Recovering from his wound, Gates returned to duty, but did not participate in any of the war's major battles. After the war, his regiment returned to England.

He did gain promotion to major in the British Army, but the end of the war in 1763 saw the British government reduce the size of their military. His regiment returned to England. However, sensing that he had no significant future in the British Army, due in part to his lack of social status, Gates resigned and sold his commission and returned to North America in 1769 to make a new start and a new life. With the assistance of his friend and comrade-in-arms from the French and Indian War, George Washington, Gates purchased a plantation in the colony of Virginia called Traveler's Rest. He became a planter, little knowing that he would soon become a soldier once again.

2. Some of Washington's generals became generals due to military acumen and experience, and others thru social status. Tell us a bit about this process, and how did Gates become a general?

News of the outbreak of the American Revolution reached Gates at his Virginia plantation in May 1775, and he immediately decided to offer his services to the Patriot cause opposing the British. The Second Continental Congress had taken the mantle of leadership in this endeavor, and George Washington served in this body as a delegate from Virginia. Recognizing Washington's military talents, the body chose him to command the army of the Patriots, and Washington accepted. He then asked Congress to appoint Gates as his adjutant. Congress did so, appointing him as the Army of the United Colonies' adjutant general with the rank of brigadier general on June 17, 1775. During his time in the British Army, Gates served in staff positions under several notable and leaders. No doubt, this experience made Gates a welcomed staff officer under Washington.

Given the rank of brigadier general, Gates helped Washington organize the force that became the Continental Army—America's first army. Even his most strident critics give Gates praise for his efficiency in this role. In his administrative and supportive role to General Washington, Gates acted swiftly creating a system of records and orders for the Continental Army. Congress promoted him to major general in June 1776. He would hold that rank until the end of the Revolutionary War.

Apparently, Gates thought very highly of himself, and believed that he could replace George Washington. Is that correct, and if so, what does history tell us about this?

As will be discussed later in more depth, Gates achieved a major success in the fall of 1777. Given command of troops in the field and tasked with stopping a British invasion from Canada, Gates not only halted the advance, but forced the invaders to surrender to him near Saratoga, New York. Although a number of his subordinates, most notably Benedict Arnold, had done the fighting that led to the capitulation, Gates received the lion's share of the credit. The defeat of the superior British Army at the Battle of Saratoga from September 19 to October 17, 1777 is considered a turning point in the Revolutionary War. The victory also persuaded France that colonial forces could prevail over the British such that in February 1778, France signed a treaty with the United States against the British giving financial and military support to the Americans.

At the same time that Gates received the surrender of the British force at Saratoga, another element of the British Army had driven George Washington from Philadelphia. In the aftermath of these two campaigns, Gates secretly worked behind the scenes to replace Washington in command of the American Army. This effort, known to history as "The Conway Cabal," did not succeed, and Gates eventually gave Washington an apology for seeking to undermine his authority.

3. What were some of his most famous battles, and what were some stories associated with him?

In the summer of 1777, British general John Burgoyne (1722–1792) invaded New York with an army of over 8,000 British and Hessian soldiers. When Congress gave Gates a field command in 1776, it had sent him to the northern frontier, and because of that Gates received orders to stop the invasion. Initially, Burgoyne met with nothing but success, advancing to within 40 miles of Albany. But a number of Gates' subordinates managed to strike blows against the British, first stalling the advance and then halting it completely.

Twice, the British attempted to break through the American defenses and push south, but both times the Americans prevailed. Knowing that he couldn't reach Albany and sensing that retreat would lead to a disastrous rout, Burgoyne chose to surrender his force to Gates. Interestingly, one regiment that surrendered to Gates was the unit that he had been assigned to when he received his commission in 1745.

When he wrote his report of the campaign, Gates downplayed the contributions made by his subordinates, making it seem that he had taken the critical actions that led to victory. This success paved the way for Congress to send him south in 1780 to counter a British campaign to conquer the colonies situated there. He found the advancing British force near Camden, South Carolina, and on August 16, 1780, he offered battle. Unfortunately, at the Battle of Camden, Gates suffered one of the worst defeats ever experienced by the American Army. Virtually every American soldier either died or surrendered. Gates was replaced by Nathaniel Greene several months later. This defeat meant that Gates would never command American troops again.

4. Apparently (and this may be wrong), he vanished after a military loss at Camden. Was he killed? Wounded in action? Body never found? This is somewhat of a mystery. Is there any agreement among historians about this?

While his men suffered death or imprisonment, Gates did not. Rather, when the battle turned against him, Gates mounted his horse and rode away. Indeed, he did not stop his ride until he had put 170 miles between him and the battlefield. Not surprisingly, when Congress learned of his behavior it removed him from command. Many wanted to court-martial Gates for his mismanagement of the battle, but his allies in Congress prevented that from happening. Amazingly, Gates eventually rejoined Washington in camp at Newburgh, New York. He became embroiled in another controversy known as the "Newburgh Conspiracy" but General Washington stifled it and nothing further came of it. Gates went on leave to care for his wife Elizabeth, who died in June 1783. When the army disbanded at the end of the Revolutionary War in November 1783, Gates's efforts were no longer required. After the war, having returned to his plantation in Virginia, Gates decided that he could not in good conscience own slaves.

As a consequence of this epiphany, Gates sold his plantation and arranged for the emancipation of his slaves, and moved to New York City in 1790. He died there in 1806. We know that his grave lies in the cemetery of New York City's Trinity Church, but the exact location of it has been lost to history.

5. What were his contributions to the American Revolution and his legacy?

Horatio Gates deserves credit for valuable service to the American cause in the first three years of the Revolutionary War. First, he helped organize what was essentially an armed mob into an effective military force. And second, although he did not play as decisive a role in the American success at Saratoga as he claimed, he did occupy the position of supreme command over that crucial effort.

However, because of his conduct during and after the Battle of Camden, Horatio Gates will always remain a largely marginal figure in the history of the war that won us our independence.

JOHN GLOVER

Saint Paul's Church National Historic Site
Image, public domain

SIX

John Glover

MARBLEHEAD MARINER ★ (1732–1797)

---

1. George Washington could not have won the American Revolution without help from his generals and their leadership. One of those generals, John Glover, is somewhat unrecognized by our history books. Where and when was he born and what do we know about his early life?

While we have only a passing knowledge of the early life of John Glover, historians believe that he was born in Salem, Massachusetts on November 5, 1732. His father made his living as a carpenter in that community. Unfortunately, Glover lost his father at the age of four. His death prompted the family to relocate to Marblehead, Massachusetts, a coastal community located a few miles from Salem. Glover would call Marblehead home for the rest of his life.

2. John Glover, before becoming a general, was known as a cordwainer (please explain!), a fisherman, a rum trader, and a merchant. What other events prepared him for the American Revolution?

Before he became involved in the American Revolution, Glover plied many trades over time becoming successful, wealthy, and influential in the Marblehad community. One of his pre-war occupations involved making shoes. English tradition drew a distinction between making shoes and repairing them. Cordwainers did the former task, while cobblers performed the latter. At some point in time, Glover then became a sailor. Even before the Pilgrims established their colony at Plymouth, Massachusetts, in 1620, English fishermen had made their way to New England to cure fish they had caught in the area known as the Grand Banks southwest of Newfoundland before taking their cargo back to England. After Massachusetts became a colony, many residents took up that endeavor, including Glover. However, not everyone involved in the maritime world made a living from fishing.

Indeed, many ships leaving New England ports carried cargoes of rum distilled in the colonies. Evidently, Glover had a stint in this trade as well. Displaying an impressive business acumen, Glover then became a merchant and ship owner in his own right. Married at 22, Glover seemed to have an idyllic life ahead of him. His fortunes, and those of Massachusetts, soon took a turn for the worse, however. A conflict known in North America as the French and Indian War began in 1754, and soon Massachusetts began to feel its effects.

A significant portion of the adult male population of Massachusetts, including Glover, served in the militia, and the colony suffered a high casualty rate. After the conflict ended, Great Britain began to impose taxes on the colonies to help pay down the debt it had incurred during the war years. These attempts at "taxation without representation" angered many Massachusetts residents, because they felt that many of their fellow residents had died during the war—literally paying with their lives. Taxing the colonists therefore seemed like a personal affront to the residents of Massachusetts who also faced other commercial restrictions by Parliament in London. Throughout the colony, communities organized committees of correspondence to coordinate a response to the British actions, and Marblehead chose Glover to serve on its local representative and later, militia commander. He was able to raise a regiment of more than 500 men that was filled by men of the militia and Marblehead seamen. An ethnically diverse group,

General Washington later, in October 1775, referred to Marblehead troops as "soldiers…bred to the sea." In a similar fashion to Glover's skill in raising a militia, Marblehead selected him to serve on a committee to enforce a non-importation policy that the First Continental Congress enacted in the fall of 1774. Over time Glover became increasingly opposed to British rule over the colonies, openly advocating independence.

By the outbreak of the American Revolution, Glover had proven that citizens could trust him with positions of authority. Not surprisingly, the people of Marblehead chose him for the position of lieutenant colonel in the militia regiment drawn from the area. When the regiment's colonel died, Glover received his rank. After the Battles of Lexington and Concord (April 19, 1775), Glover marched his unit to Boston, and became part of the army that George Washington would soon command.

3. In a sense, John Glover's ship, a schonner named *Hannah*, was the precursor of the United States Navy, or am I off on this?

After Washington and Glover became acquainted, Washington learned of Glover's nautical background, including the fact that Glover owned a vessel. Named after Glover's daughter, the *Hannah* had a shallow draft, making it a perfect craft to operate in coastal waters. For that reason, Washington requisitioned the vessel (August 24, 1775), and sent it out under the command of an experienced shipmaster serving in Glover's regiment (Nicholson Broughton, 1724–1728). The success that the *Hannah* enjoyed in capturing British vessels prompted the Continental Congress to create a navy under its direct command. Because of this, many historians regard the *Hannah* as the first ship in the United States Navy in that it was the first armed vessel to sail under Continental control and pay. (The Continental Navy was authorized by Congress on October 13, 1775, after *Hannah*'s acquisition.)

Hannah's mission was to aid against the British siege of Boston (April 19, 1775–March 17, 1776) by capturing provisions ships headed to Boston Harbor with supplies for the British. *Hannah* had two engagements. The first was with the sloop Unity on September 7, 1775 and *Hannah* readily defeated Unity and captured British stores and provisions. The second conflict, on October 10, 1775 with the sloop *Nautilus*, resulted in *Hannah* being run aground off Beverly, Massachusetts. However, townspeople on shore successfully engaged *Nautilus* and *Hannah* was saved from capture or destruction. Soon thereafter, with better ships being made available and built, *Hannah* was decommissioned.

4. Many men died in the American Revolution, yet Glover, due to his nautical skills and leadership, apparently saved many men. What do we know about this?

In August of 1776, Washington suffered a potentially crushing loss at the Battle of Long Island. It seemed likely that the British would capture the remnants of Washington's army, but he decided that he would try to save his force through a desperate measure. Washington called upon Glover to have the men in his 14th Regiment, the "Marblehead Mariners," gather up all the boats that they could find and use them to ferry his army of 9,000 troops, horses, artillery, and equipment from Brooklyn across the East River to Manhattan.

Using muffled oars and benefiting from a fog that blanketed the region the night of the evacuation, Glover and his men managed to ferry virtually all of Washington's men to safety. The mission was accomplished in nine hours. Much as the evacuation of Dunkirk saved the British cause during World War II, the actions of Glover and his men allowed the American cause to live on during the American Revolution.

5. Did Glover's "Marblehead Mariners" make any other direct contributions to the Patriot cause?

After Washington evacuated New York in the fall of 1776, he retreated across New Jersey to the Delaware River. There, Glover's regiment (nicknamed the "Marblehead Mariners" by author George Billias in his 1960 book on Glover) ferried Washington's army to safety on the Pennsylvania shore. Recognizing the dire situation that he faced, Washington decided to risk everything on an effort to make a surprise attack on a Hessian force guarding Trenton. To accomplish this task, Washington wanted to assault the Hessians at daybreak on the morning of December 26, 1776. That timing would require his army to cross the Delaware in the middle of the night.

As he had four months earlier, Washington called upon Glover to effect a nighttime crossing. While Glover's men had fog to contend with in August, in December they faced ice floes and a snowstorm on the Delaware River. But once again, Glover's men overcame the elements, and successfully transferred Washington's men to the New Jersey shore. Glover's regiment then took up arms, and joined Washington's main force for its successful attack Christmas Day attack on the Hessians. Glover and his men then transported nearly 1000 Hessian prisoners back across the river on the same day. This event is the famous crossing of the Delaware by Washington commemorated in the magnificent 1851 painting by Emanuel Leutze titled simply *Washington Crossing the Delaware.*

6. Apparently, Glove had to return home to tend to his sick wife, Hannah, but a personal appeal from none other than George Washington encouraged him to return to duty. What are the details of this?

During the Revolutionary War, most American volunteers' term of service ended on December 31st of the year of their enlistment. For that reason, in early January of 1777 virtually all of Glover's regiment—Glover among them—opted to return to civilian life after their enlistments expired. Even the offer of a promotion to the rank of brigadier general could not convince Glover to remain with the Continental Army. At that moment, Washington interjected himself into the situation, personally asking Glover to return to military service.

Moved by his entreaty, Glover did return, and accepted promotion to the rank of brigadier general and command of a brigade. Assigned to the army commanded by Horatio Gates, Glover's brigade participated in the campaign that resulted in the capture of the British-Hessian army commanded by General John Burgoyne. He then became a part of an expedition in 1778 to regain control of Providence, Rhode Island. Although the campaign failed, the men under his command fought well. In this manner, Glover certainly proved worthy of the confidence that Washington showed in him by asking that he rejoin the army with the rank of brigadier general.

7. Part of his "tour of duty" during the revolution was guarding the Hudson River. Why was this so important, and what were his duties?

Throughout the American Revolution, the British recognized that gaining control of the Lake Champlain-Hudson River corridor from Canada to New York City could seriously jeopardize the hopes of the United States to achieve independence. Conversely, the leaders of the fledgling nation recognized the importance of keeping that route in its hands. For that reason, Washington had to commit valuable resources to the corridor, Glover among them. Although he saw no action during his time there, Glover nonetheless helped the American cause by providing a deterrent to any thoughts that the British might have had about attempting another campaign like Burgoyne's.

8. In closing, there is a statue of John Glover in Massachusetts honoring his life, his sacrifices and his accomplishments. Where is the statue? And in closing, what can you say to honor this great man who gave so much to Washington and the American Revolution?

Visitors to Boston can find a statue honoring John Glover on Commonwealth Avenue. The state of Vermont has a city named after him. Even though he served in the Continental Army, the Navy recognized his nautical contributions to our nation by naming a frigate after him—the USS *Glover* (AGDE-1, later FF-1098). Outside of these commemorations, Glover's name is scarcely remembered today. He died in 1797 was buried in the Old Burial Hill Cemetery, Marblehead, Massachusetts.

Ironically, Elbridge Gerry (1744–1814), a person that served with him on a committee of correspondence prior to the American Revolution, will live forever in the national consciousness through his association with the term "Gerrymandering." An officer who helped save the American cause on two separate occasions through his nautical abilities deserves to be much better known by the American people.

NATHANIEL GREENE

By Charles Wilson Peale, 1783
Independence Hall, Philadelphia (INDE 14055)
Image, public domain

SEVEN

Nathaniel Greene

SOUTHERN STRATEGIST ★ (1742–1786)

---

1. As we study the American Revolution, and the generals and heroes who assisted George Washington, I see at least two trends. Some of the generals were Quakers, and some were Masons, and of course, others weren't either or were a combination socially and religiously. We will discuss each of these issues as we traverse our study of Washington's Generals. Today, we look at Nathanael Greene, who was born a Quaker and walked with a severe pronounced limp. What do we know about his childhood?

While most history books refer to him as "Nathaniel" Greene, his parents actually named him Nathanael Greene. Interestingly, his father had the same name, but did not name his son Nathanael Greene, Jr. even though "Junior" had been used in names since the 1500s. The Nathanael Greene that would become an American general during the Revolutionary War was born on August 7, 1742 on a farm located near Warwick, Rhode Island.

A well-to-do man, the senior Nathanael Greene could have easily afforded to send his son to school, but chose not to. This decision stemmed from the fact that the senior Greene belonged to the Society of Friends (better known to most people as the Quakers), and his branch of the faith believed that members should not pursue education simply for the sake of learning. Instead, Greene grew up learning the business of his father's mills and foundry.

For that reason, the younger Nathanael Greene had to teach himself subjects such as mathematics. He also began to read history books, especially ones that focused on military matters. This put him further at odds with his faith, because Quakers practiced pacifism. Greene's interest in the military inspired him in 1774 to enlist in a Rhode Island militia unit. Because of his prominent position in society by that time, Greene assumed that he would become an officer in that unit, but he did not receive such an appointment or commission.

Although given no reason at the time, Greene felt that the slight might have resulted from the fact that he suffered from what we would call today a frozen knee. This condition forced him to limp and this probably affected the way that others perceived him. Rather than leave the militia because of this, Greene simply agreed to serve as a private. However, his status soon changed. After the Battles of Lexington and Concord (April 19, 1775), the Rhode Island legislature authorized the formation of an "Army of Observation," and ordered it to proceed to Boston.

For reasons which remain unclear, the legislature made Greene the major general of its militia. It is unusual in American history for individuals to go from private to general as did Greene (though not unheard of). In a similar fashion, few Quakers gave up their pacifist beliefs to fight for the American cause, as Greene did. He is a very unique figure in our nation's history.

2. Washington apparently thought highly of Greene, giving him control of the city of Boston. How did this come about?

Greene and his Rhode Islanders reached Boston in June of 1775. By then, the Second Continental Congress had appointed George

Washington to command the Continental Army. When Washington reached Boston, Greene met him to offer his greetings. Washington sized up his subordinate, quickly recognizing that Greene's independent studies had given him a firm grasp of military matters. Furthermore, he appreciated Greene's lack of pretension. As the Americans began a siege of Boston, Washington appreciation of Greene's talents as a military leader grew with every passing month.

Obviously, others saw the same qualities in Greene, as even before Washington assumed command the Second Continental Congress had given Greene the rank of brigadier general in the Continental Army. Washington and Greene discussed plans for driving the British out of Boston through an assault, but after the Americans captured artillery pieces at Fort Ticonderoga in May, 1775 by Ethan Allen and his "Green Mountain Boys" militia, Washington used the weapons to force a British evacuation in March of 1776. Washington then gave Greene command of the city, but soon called upon him to journey to New York to help defend that city. Greene experienced some defeats in the New York Campaign in that he lost Forts Washington and Lee in November 1776 in the struggle to defend the Hudson River. However, he determined to recover from those losses and acted swiftly at the Battle of Trenton (December 1776) and the Battle of Princeton (January 1777) and was victorious.

3. It is true that George Washington had indicated that if he were to be killed in battle, he wished Nathaniel Greene to take over in his place?

While certainly not a fatalist, George Washington recognized that any number of factors could lead to his death. Accordingly, he decided to officially designate a successor should he die before the United States had won its independence. Washington had a number of capable subordinates to choose from, but apparently, he never seriously considered anyone other than Greene. Fortunately, the nation never needed this contingency plan, but it still speaks well to the confidence that Washington placed in Greene.

4. Greene was also "Quartermaster General" of the Continental Army. When was that and what exactly were his duties at that time?

After Washington went into winter quarters at Valley Forge in the late fall of 1777, he recognized that his army lacked the supplies that it would need for a winter encampment. This coincided with the resignation of Quartermaster General Thomas Mifflin. Even though Greene had served in a purely combat role up to that point in time, Washington felt that he had the capabilities necessary to keep the Continental Army supplied. For that reason, after some months Congress appointed Greene Quartermaster General with the rank of major general. This position entailed procuring the food, clothing, and shelter necessary for an army to function, and required superior administrative skills.

Though reluctant to accept the appointment, he did so with Washington's urging and with the understanding and approval that his role would not prohibit him from leading troops in combat in the field. Greene proved up to the task, in spite of the fact that he had limited financial resources with which to work. The soldiers would suffer from hunger that winter, but most historians believed that they would have suffered far more than they did with someone other than Greene serving with Washington during the harsh winter. At Valley Forge, Washington asked him to bring some order to the Quartermaster department and he did so reluctantly, commenting "no one remembers the Quartermaster." He served as Quartemaster General from 1778–1780.

5. Some of Greene's greatest successes were in the South against British leader Lord Cornwallis (1738–1805). How was Greene so militarily effective in the South?

During the first five years of the Revolutionary War, Nathanael Greene primarily served in a combat capacity, and proved himself a valorous and capable commander. He made mistakes occasionally, but he always learned from them. For that reason, George Washington gave him a crucial independent command in 1780. It was a dark year for the Patriot cause. That year had seen the British capture Charleston, South Carolina, and the beginning of a campaign to crush the American cause in the South. The British believed that the strongest loyalty to the crown was in the southern colonies and that victory there would swift. In response, Congress sent Major General Horatio Gates (1727–1806) to stabilize the situation, but he had suffered a disastrous defeat at the Battle of Camden (August 16, 1780). Congress then asked Washington to appoint a successor, and he chose Greene.

When Greene arrived in South Carolina in December of 1780, he immediately adopted the seemingly suicidal strategy of dividing his small force. Greene commanded one wing, and gave the other to General Daniel Morgan (1736–1802). In actuality, this rash move was a stroke of genius, as it forced the British to then divide their forces to deal with the two groups of Americans. Hoping to crush the smaller American forces, the British failed miserably.

First, Morgan succeeded in decisively defeating the British sent to confront him at the Battle of Cowpens in January 1781. The other British operation fared no better, as Greene inflicted heavy casualties on them at the Battle of Guilford Court House in March 1781 (even though he had technically lost the battle, having chosen to abandon the field of battle after bloodying the enemy).

Unable to crush Greene, the British chose instead to march the majority of their soldiers north into Virginia. It was this army that Washington would force to surrender at Yorktown in October 1781, an act that effectively ended the Revolutionary War. Greene's southern campaign strategy and operations thus led to the ultimate American triumph, even though he personally had not won a battle in that region. However, his leadership in the South continued the rise of his stature such that he would be remembered as one of the finest strategists and generals of the American Revolution.

6. What have I neglected to ask about this famous American general who so ably assisted George Washington?

Ironically, even after his death Greene played a role in American history. In gratitude for his contributions to the American cause, the state of Georgia had given him a plantation. He and his wife moved there after the war. When Greene died in 1786, his widow chose to remain in Georgia, and soon met a young man from Connecticut who had come to Georgia as a tutor. In discussions with him, Greene's widow learned that the young man had an inventive mind, and she allowed him to live at her plantation while he worked on bringing his dreams to life. That young man was Eli Whitney, who would invent the cotton gin while living at the plantation. Nathanael Greene thus influenced his nation perhaps even more after death than he had while serving in the American Revolution.

JOHANN DE KALB

Replica of original
By Charles Wilson Peale, 1781-1782
Independence Hall, Philadelphia (INDE 14084)
Image, public domain

EIGHT

Baron Johann de Kalb

THE "OTHER" GERMAN GENERAL ★ (1721–1780)

1. Americans owe so much to so many people who came from foreign lands to assist in the American Revolution. Baron Johann de Kalb was once such soldier and leader. Where exactly was he born and what was his nationality?

Unlike most American generals, certain details about the early life of the individual known in American history as Baron Johann de Kalb remain shrouded in mystery. Historians agree that he was born in June 1721 in the Bavarian village of Hüttendorf, but differ on the type of family into which he was born. Some accounts, for example, state that his parents had the status of peasants, while some assert that his parents belonged to the nobility in the principality of Bayreuth.

Judging by the fact that he spoke French and English in addition to his native German tongue, it seems likely that his parents must have had the financial wherewithal to send him to more than just a rudimentary local school. It seems fair to say, then, that Johann Kalb—his actual name, according to most historians—came from a comfortably situated family, if not a noble one. In 1738, Kalb left Hüttendorf, never to return.

2. Regarding de Kalb's prior military experience and training, what was he involved with?

Although he was born in a German principality, Johann Kalb chose to seek a career in the French Army. Even though French and Germanic forces fought each other from the seventeenth century onward, France proved able during the eighteenth century to recruit whole regiments from what would later become Germany. In 1743, Johann Kalb received a commission as a lieutenant in the Loewendal Regiment, one officered exclusively by German-born officers. Interestingly, historians know next to nothing about the period in Kalb's life between his leaving Hüttendorf and his becoming an officer in the Loewendal Regiment. One thing that definitely did happen during those years involves Kalb's name.

From 1743 on, he referred to himself as "Baron de Kalb," meaning that he had assumed a title of nobility. Undoubtedly he did so to present himself as suitable officer material. Once he became an officer, de Kalb had an immediate baptism of fire. Three years before de Kalb entered the military, France had become involved in a conflict known in Europe as the War of the Austrian Succession, and he joined his regiment while it was engaged in combat in Flanders. De Kalb served the French cause ably, earning a promotion to the rank of captain by the end of the war. After the warring parties negotiated a settlement to the conflict in 1748, de Kalb chose to remain in the army.

Promoted to major, de Kalb saw action once again during the Seven Years' War, earning a further promotion to lieutenant colonel. When that war ended, de Kalb resigned his commission and married the fifteen-year-old daughter of a wealthy French fabric merchant. Soon after their marriage, his wife inherited the family fortune. De Kalb could have simply lived the rest of his life as a man of leisure, but at the age of 55 he decided to take up arms again and join the American cause during the Revolutionary War.

3. He apparently worked with both the Marquis de Lafayette and George Washington. What do we know about his relationship with them?

Although only a teenager at the start of the American Revolution, the Marquis de Lafayette wanted to help the new nation achieve its independence. Because he had no military experience, however, it seemed unlikely that he would get the opportunity to offer his services to the Patriot cause. Fortunately for Lafayette, a French nobleman had introduced him to de Kalb. Although vastly different in virtually every respect, the two became good friends.

Since de Kalb also longed to fight for the fledgling nation, the two decided that they would jointly offer their services. For that reason, they approached Silas Deane, the American representative to France. Deane had granted appointments to the rank of general for a number of Europeans, and in 1777 Lafayette and de Kalb made a similar request. Happy to oblige them, Deane gave them both appointments to the rank of major general. On April 20, 1777, the two boarded a French ship bound for the United States. Once they arrived at their destination, they found out that Deane had exceeded his authority in giving them their appointments. As it turned out, Washington did not have enough soldiers in his army to create units that major generals commanded by virtue of their rank.

Lafayette and de Kalb reacted quite differently to the news that they would not receive their appointments. On the one hand, the young Frenchman said that he would serve as an unpaid volunteer, and joined Washington's army. De Kalb, on the other hand, said that he would return to France, and went on to say that he expected Congress to pay his expenses for his journey to and from America. Proving true to his word, de Kalb immediately started making his way back to the coast.

For reasons that remain unclear, Congress chose to confirm both his and Lafayette's appointments. By the time de Kalb reached

Washington's army, the British had taken possession of Philadelphia. When he arrived, he received an invitation to a meeting of all the generals serving in Washington's army. At that gathering, Washington informed them that he hoped to launch an assault on Philadelphia, much as he had done at Trenton. De Kalb and a majority of the generals strongly advised against that course of action, pointing out that the British had built stout fortifications to defend the city. As he often did, Washington deferred to the opinion of the majority, and did not follow through on his plan.

4. He spent some time at Valley Forge. What are the details?

After Washington gave up his plan to attack the British in Philadelphia, he reluctantly decided to make camp for the winter of 1777–1778. Wanting to keep a close eye on the British, he chose a nearby location known as Valley Forge. Here again, de Kalb differed with Washington, as he believed that the army would find it hard to keep itself supplied in the remote location. Although one could argue that Washington would have had trouble keeping his army fed in any location, the hardships that his troops faced certainly bore out the accuracy of de Kalb's opinion regarding the disadvantages of wintering at Valley Forge, though de Kalb remained at Valley Forge during the harsh winter.

5. Often military individuals give each other the highest regard and respect. What do we know about what transpired in this regard after the Battle of Camden?

When the British evacuated Philadelphia in the spring of 1778, Washington decided to attack them as they retreated through the New Jersey countryside. This resulted in the Battle of Monmouth on June 28. During that engagement, de Kalb's men participated in a stalwart stand by the right wing of the Continental Army that helped blunt attacks made by British General Charles Cornwallis, and allowed Washington to counterattack late in the day. Impressed with de Kalb's leadership, Washington gave him an important assignment in the spring of 1780. Learning that the British had sent a detachment of over 2,000 men to join operations in South Carolina, Washington ordered de Kalb to take 1,400 Continental soldiers to reinforce General Benjamin Lincoln in Charleston, South Carolina.

While on the route southward, de Kalb received news that Lincoln had surrendered the city, and all the soldiers under his command, in early May. Receiving no orders to the contrary, de Kalb continued to march toward South Carolina. While camped in North Carolina, he learned that Congress had sent General Horatio Gates to take command of American forces in the South. Once Gates arrived, he informed de Kalb that he planned to move on a British outpost at Camden, South Carolina. Picking up militia units along the way, Gates and de Kalb reached their destination by mid-August. Even though the march had tired his men, Gates ordered them on a night march on August 15 to establish a position that would threaten the British garrison in Camden.

Much to their surprise, the advancing Americans ran into British soldiers moving in their direction. It turned out that the British had learned of Gates' arrival in South Carolina, and had sent a force to intercept him before he reached Camden. The Americans had run into the vanguard of that force on the night of the 15th. Alerted to the presence of the British, Gates drew up a plan of battle for the following day. Placing his militia on his left wing with orders to hold their positions, Gates ordered de Kalb to use all of his soldiers to launch an attack from the right wing.

At the start of the battle, de Kalb's men moved forward, but at the same time a British force attacked the American militia. Every militia unit save one quickly fled the battlefield, leaving de Kalb and his men to fend for themselves. Had the militia held their positions, Gates might have won the battle, as de Kalb's attack came close to breaking the British position, but after the collapse of the left wing of Gates' army the British turned their full attention to de Kalb's men. His outnumbered force fought valiantly, urged on by de Kalb. Soon, however, the superior numbers proved overwhelming, and the British overwhelmed the brave stand of de Kalb and his men. After receiving three wounds from musket balls and eight wounds from bayonets, de Kalb finally collapsed. Taken prisoner by the British, de Kalb lived for three days before finally succumbing to his wounds.

6. Where is de Kalb buried, and what were some of the things said about this heroic military leader?

After he died, de Kalb was originally buried in a nearby cemetery. In 1825, his body was moved to the Bethesda Presbyterian Church cemetery in Camden. A monument to his honor was placed there, with Lafayette laying the cornerstone for his esteemed former comrade. A plaque was placed on the monument, stating "Here lies the remains of Baron De Kalb, a German by birth, but in principle, a citizen of the world." This seems a fitting tribute for a man who made our nation's fight for its independence his own.

7. What else should we know about this singular individual?

Six states have counties named for de Kalb, making him the third most-honored general of the American Revolution. Obviously, Americans of an earlier age greatly respected a man who refused to surrender despite suffering unimaginable pain on the field of honor that day at Camden.

HENRY KNOX

By Charles Wilson Peale, c. 1784
Independence Hall, Philadelphia (INDE 14081)
Image, public domain

NINE

Henry Knox

BOOKISH AND BOLD ★ (1750–1806)

---

1. Henry Knox was a true American hero and for a while, the youngest general in the American Revolution. What do we know about his early years?

Henry Knox was born on October 25, 1750, in Boston, Massachusetts. His father owned a shipbuilding company in that city, and during Henry Knox's early years his father's business thrived. His father suffered financial difficulties during the French and Indian War, however, and the elder Knox felt compelled to seek new business opportunities in the Caribbean. Unfortunately, he died there in 1762.

At the time, Henry Knox attended the Boston Latin School, but his father's death forced him to drop out and seek employment to help support his family. Knox apprenticed himself to a local bookbinder and bookseller named Nicholas Bowes, and Knox soon proved well suited for that occupation as well as selling books. Bowes took a liking to Knox, and gave him the privilege of taking home any book in the store to read. While he read books on a variety of subjects, Knox had a preference for works that dealt with military history and tactics. In addition, Knox read a number of works on engineering, artillery, and ordnance. He also taught himself to read French so that he would be able to read military titles that not yet been translated into English.

In 1768, at the age of 18, Knox joined a joined the artillery company known as the Company of Artillery or "The Train"—men from the South End of Boston under the command of Major Adino Paddock (a Loyalist), a chair maker. It was here where Knox got his first formal artillery training. (In colonial New England, Massachusetts law required nearly all males between 16 and 60 to drill with their town militia units in order to provide protection if needed.)

During the Revolutionary War, Knox would put the knowledge that he gained from reading those books to good use. In 1771, at the age of 21 and after working for Bowes for nearly a decade, Knox decided to open his own bookstore, which he named The London Bookstore. Not surprisingly, his bookstore contained many works on military history and attracted many British officers as well as prominent citizens of Boston such as John Adams. Initially, Knox's business did well, but the British decision to close the port of Boston to trade in 1774 caused him great financial distress.

In 1772, Knox's interest in all things military led him to join a new militia unit, the Boston Grenadier Corps under command of Captain Joespeh Pierce. Knox was made second in command

By that time Knox had already become identified with opposition to British colonial policies through his membership in the Sons of Liberty, and British actions in 1774 no doubt drove him further in that direction. When hostilities between Great Britain and the colonies broke out in 1775, Knox thus stood poised to join the Patriot cause.

2. Henry Knox was witness to the Boston Massacre and allegedly took part in the trial against the British soldiers. How might this have shaped his view of things?

Tensions between the colonies and Great Britain had developed soon after the end of the French and Indian War, largely over the British attempts to tax the inhabitants of North America. Actions by the colonists included the destruction of governmental property, and in response the British had sent troops to the cities where the vandalism had taken place. In Boston, many of the citizens took to the streets to protest the presence of the British soldiers, and on the night of March 5, 1770, this animosity led to British soldiers firing on a group of Bostonians. Five civilians died, and the British soldiers responsible for the killings soon found themselves on trial for murder.

Henry Knox testified at the trial, stating that he had seen the incident take occur. He claimed that he had tried to convince the British soldiers to stand down, a suggestion that they failed to act on. This account appears plausible, as his biographer found evidence that Knox had belonged to what we would call today a street gang from the neighborhood where the Boston Massacre took place. It's logical to assume that witnessing the British action further helped to politicize Knox.

3. The Boston Tea Party, the Battles of Lexington and Concord led Knox to join the American Revolution. How did these events impact Knox?

Knox became an ardent supporter of the Patriot cause and eventually, a major military leader in it. No one knows whether Henry Knox actively participated in the Boston Tea Party or not, but historians agree that Knox did play a role in the steps that led up to that event. When a ship carrying tea arrived in Boston in November of 1773, a group of citizens went to the wharf to insure that no one unloaded the cargo. This contingent included Henry Knox. As previously noted, Knox became even more opposed to British policies after the passage of the Boston Port Bill in 1774—an Act to punish Boston and Bostonians for the events of the Boston Tea Party (December 16, 1773) by closing the port to commerce. Many other residents of Massachusetts shared his opinion, they began to engage in military training. Eventually, this led to the Battles at Lexington and Concord in April 1775.

Although Knox had by that time participated a militia unit in Boston, he did not participate in those battles, but in their aftermath he joined the force that had driven the British back into Boston. He closed his book shop and left Boston to join those mobilizing across the Charles River in Cambridge in the aftermath of the Battles of Lexington and Concord. When he did so, his wife, the former Lucy Flucker, who he married in June 1774 rode with him, hiding his sword in her cape.

Recognizing his grasp of military matters, Artemas Ward (the commander of the colonial force) asked Knox to help build a series of fortifications and gun emplacements around Boston to help keep the British contained inside Boston. Knox thus became a part of the Patriot forces in the spring of 1775, although he would not receive an official commission until months later.

When George Washington assumed command of the Continental Army outside of Boston, he toured the fortifications that Knox had built and met with him personally. Impressed with Knox, Washington soon recommended that he receive a commission as colonel of artillery. Congress soon agreed, and Knox began a career in the military that would last until he resigned in 1784.

4. At the Battle of Bunker Hill (June 17, 1775), what was Knox's involvement?

From an early age, Knox had read extensively about military history. Then, when he became legally able, Knox enlisted at the age of 18 in a Massachusetts militia artillery unit. Because of this, Artemas Ward would eventually ask Knox to command his artillery units deployed outside Boston. But this happened after, rather than before, the Battle of Bunker Hill.

At that engagement, a short two months after the Battles of Lexington and Concord, American artillery had played virtually no role, because the soldiers manning the cannons fled the field of battle as the British began their assault. Ward relieved the artillery commander because of his poor battlefield performance, and Knox (who had fought at the Battle of Bunker Hill as a volunteer) would eventually receive that position.

5. What were some of his greatest contributions?

Knox gave invaluable service to the Patriot cause right up to the end of the Revolutionary War, but his crowning achievement came relatively early in the conflict. After the Battle of Bunker Hill, it became obvious that the British Army did not have sufficient strength to break Washington's siege. Both sides also recognized that Washington had no realistic chance of taking Boston through a frontal assault. Washington recognized that heavy artillery could force the British to abandon Boston, but when he assumed command of colonial forces in the summer of 1775 he learned that his army had only a few light artillery pieces. Soon, however, the situation changed because of the capture of Fort Ticonderoga in New York by Patriot forces of Ethan Allen and Benedict Arnold.

Most historians believe that Henry Knox recognized that if the Continental Army could move the cannon captured there to Boston, Washington could quickly force the British out of that city. Knox created a plan and presented it to General Washington. Washington gave his approval, and in early December 1775, Knox journeyed to Fort Ticonderoga to make arrangements for this undertaking.

Remarkably, within six weeks Knox proved able to move 59 artillery pieces with a combined weight of 60 tons over 300 miles through a tractless wilderness in the dead of winter. Most were "12-pounder" or "18-pounder" pieces, referring to the weight of the cannonballs they fired. However, there was one "24-pounder" nicknamed "Old Sow" that weighed more than 5,000 pounds. Known as the "noble train of artillery," Knox and his men accomplished their march through the Bershire Mountains with the assistance of teams of oxen in order to bring the artillery pieces to Boston

When Knox and his train of artillery arrived in Boston, Washington decided to deploy them on the Dorcester Heights overlooking the city. Once the British saw the cannon on that prominent location and elevation, they recognized that they had to evacuate Boston. They did so in mid-March 1776, and never returned. As they departed after an eleven-month siege, General Washington was there on the heights watching it all—thanks to 25-year old Henry Knox. Every year, "Evacuation Day," March 17, is celebrated as an official holiday in Boston—and has been since 1901. History offers few equals to the logistical feat that Knox accomplished.

Later in the war Knox oversaw the logistics for Washington's infamous crossing of the Delaware River in December 1776. He also participated in the victorious October 1781 siege at Yorktown.

6. In his later years, how did he serve his country after the war?

Knox remained in the army until 1784, and then resigned his commission. He then became involved in an effort to amass a vast amount of land of which his in-laws (who had remained loyal to King George III throughout the Revolutionary War) had abandoned their ownership claims. Instead of becoming a gentleman farmer, however, Knox soon came back to public service.

After the resignation of Benjamin Lincoln, our nation's first Secretary of War, Congress asked Knox to replace him. Assuming that office in 1785, Knox continued to serve in that capacity even after the nation adopted the Constitution in 1788. As Secretary of War, Knox helped organize both a permanent American Army in 1785 and a permanent

American Navy in 1794. In this manner, Knox gave invaluable service to the nation both in war and in peace.

7. What have I forgotten to ask about one of Washington's finest generals?

As we previously have seen, Knox had no formal military training whatsoever before offering his services to the Patriot cause in 1775. Six years later, after he had commanded American artillery at the Siege of Yorktown, the French liaison to the American Army referred to Knox as a "military genius." Knox thus exhibited perhaps the steepest learning curve in the history of our nation's military. Without question, the bookseller from Boston is the Father of American Artillery.

THADDEUS KOSCIUSZKO

By Julian Rys, after unidentified source, 1897
NPS/Independence National Historical Park
Image, public domain

TEN

Thaddeus Kosciuszko

POLISH PATRIOT ★ (1746–1817)

---

1. Few people have heard of or know much about General Thaddeus Kosciuszko. Who was he?

Thaddeus Kosciuszko was born in February of 1746 near what is now Kosava in the Republic of Belarus. His father held a title of nobility, and served as an officer in the Polish-Lithuanian Commonwealth Army. Interestingly, his parents had him baptized in both the Roman Catholic and Eastern Orthodox Churches. Kosciuszko family sent him to school in the Ukrainian town of Lyubeshiv, but his father's death forced him to abandon his studies in 1758. For the next seven years, Kosciuszko helped his mother manage the family's estate, but in 1765 another educational opportunity presented itself to him.

In that year, the King of Poland started a program at the university in Warsaw to train officers for his army, and Kosciuszko received an appointment. Graduating in 1766, Kosciuszko then became a junior officer in the Polish Army. Instead of joining Polish forces in the field, Kosciuszko received orders to remain at the institute and become an instructor. Two years later, a rebellion broke out in Poland, and Kosciuszko soon found himself conflicted about which side to support.

On the one hand, as an army officer he felt loyalty to the king. But on the other hand, one of his brothers had joined the rebellion. Unable to decide which faction to fight for, he chose instead to leave the country. He asked the king for a scholarship to study in France, and in 1769 the monarch granted his request. For the next five years, Kosciuszko studied art, architecture, and military science in Paris. When he returned to his homeland, he became a tutor to the daughter of a provincial governor.

The two fell in love, but the governor made it clear that he did not approve of the romance. Believing that they would never receive his blessing, the two attempted to elope, but individuals working for the governor thwarted their plan. After receiving a beating, Kosciuszko decided to leave Poland. Lacking any other option, he chose to return to Paris in 1775. Unbeknownst to him at the time, that decision would put him on a path to becoming a part of the American Revolution.

2. What do we know about his reasons for coming to American to help the colonists with the American Revolution?

When Kosciuszko arrived in Paris, he learned that 13 British colonies in North America had initiated a rebellion against their mother country. Although the French government adopted a policy of neutrality, many French citizens openly supported the rebellious colonies. As he circulated through Parisian society, Kosciuszko expressed similar sentiments.

This may have resulted from a hostility to Absolutist Monarchy stemming from the treatment of his native Poland by Russia and Austria, or he could have developed a liberal outlook during his study in Paris. Soon, Kosciuszko chose to support the rebellion directly, deciding in June of 1776 to sail to North America and offer his services to the insurgents.

3. As a historian, what do you see as his major contributions?

Upon his arrival in North America, Kosciuszko found that the situation there had changed dramatically. When he sailed in June of 1776, the 13 colonies had used military force to oppose the British, but they had not taken the irreversible step of declaring independence. As every American knows, they did so on July 4. After his boat landed, Kosciuszko thus had to decide whether to support a movement to create a new nation or not.

Without hesitation, Kosciuszko wrote a letter to the Second Continental Congress seeking a commission. Congress granted his request, giving him the responsibility of constructing forts to defend Philadelphia. At first he served as a volunteer, but in October Congress commissioned him a colonel in the American Army. Sent north to aid General Horatio Gates in his efforts to stop a British invasion that had originated in Canada, Kosciuszko immediately proved his worth by giving Gates sound advice on building fortifications to impede the enemy's progress.

In 1778, Kosciuszko helped the American Army improve the fortifications at West Point, New York, the current site of the nation's military academy. Two years later, he received orders to join the American force attempting to defend the southern colonies. Initially he served as an engineer, but as the war progressed he was given command of a combat force. In that capacity, he fought a number of battles in South Carolina in 1781–1782. Because of his meritorious service, Congress promoted him to brigadier general, and gave him gifts of land and money.

4. Without television, Internet, or social media during his era, how exactly did Kosciuszko learn about the American Revolution?

Although the American Revolution started simply as a rebellion by 13 colonies against Great Britain, news of this event quickly piqued the interest of Europe. This stemmed from the fact that many of that continent's nations had fought either with or against Great Britain in a series of wars from 1689 to 1763. Europeans thus either wished the

British well in their efforts to suppress the rebellion, or hoped that the colonies would cause their former enemy great harm.

Because France had been Great Britain's main foe, the French had the greatest interest in the course of the American Revolution—so much so that they went to war with Great Britain in 1778 to help the United States in its effort to secure independence. Because of this interest, Kosciuszko found an abundance of information in Paris about the American Revolution after he arrived there.

5. How is he regarded currently in Poland?

At the end of the American Revolution, Kosciuszko returned to Poland. For a few years, he helped operate his family's estate, but in 1789 he agreed to accept a commission in the Polish Army. Three years later, a war with Russia broke out. During the conflict, Kosciuszko won a number of victories, but when Prussia entered the conflict the Poles suffered a series of defeats. In 1795, with Kosciuszko wounded and in captivity, Poland found itself unable to continue to resist. They capitulated, and Russia and Prussia then divided the country between themselves. For the rest of his life, Kosciuszko devoted his efforts to re-establishing an independent Poland. Because of this, he became a hero to his people. He thus holds an esteemed position in Poland (which finally came back into existence at the end of the First World War) to this day.

6. What have I forgotten to ask about this famous contributor to our freedom?

Many reminders of Kosciuszko's service to the American cause remain to this day. Perhaps the most interesting can be found in Philadelphia. After Poland ceased to exist, Kosciuszko had returned to the United States in 1797, taking up residence in Philadelphia. The building that he lived in became a National Historic Place in

1970, and in 1972 it became the Thaddeus Kosciuszko National Monument. This has the distinction of being the smallest U.S. national monument in existence.

GILBERT DU MOTIER, MARQUIS DE LAFAYETTE

By Charles Wilson Peale, 1779–1780
Independence Hall, Philadelphia (INDE 14080)
Image, public domain

ELEVEN

Lafayette, Gilbert du Motier, Marquis de

A FRIEND WHEN NEEDED ★ (1757–1834)

1. In this series of interviews, we are examining the many individuals who aided and abetted the colonists in their drive for Independence. We now consider Marquis de Lafayette. Where was he born, and what do we know about his background?

Gilbert du Motier, Marquis de Lafayette, was born into a wealthy and noble French family on September 6, 1757, in Chavaniac, France. His family had a distinguished heritage, as one of his ancestors had served as an officer in Joan of Arc's 1429 campaign to capture Orléans. At the time of his birth, Lafayette's father held the rank of colonel in the French Army. Two years later, Lafayette's father lost his life at the Battle of Minden during the Seven Years' War.

Upon his father's death, Lafayette inherited his father's title of Marquis and Lord of Chavaniac. He lived on his family's estate until he turned 11, when he moved to Paris and enrolled in the Collège du Plessis (known as the University of Paris or the Sorbonne). Soon thereafter, his great-grandfather arranged for Lafayette to begin training to become an officer in the Musketeers of the Guard, a military branch of the royal guard known as the Maison du Roi. Commissioned at the age of 14, Lafayette would transfer to a dragoon regiment two years later.

At the age of 17, Lafayette married Marie Adrienne Francoise. When he turned 18, Lafayette's superiors promoted him to the rank of captain. Soon, however, Lafayette decided to leave the French Army and offer his services to the United States.

2. How did Lafayette first learn about the American Revolution? Obviously, there was no internet back in 1776.

At one time, France had held the claim to a vast expanse of land in North America, but a series of wars against Great Britain had gradually cost France much of that territory. Because of this, many French citizens harbored feelings of resentment against their rivals. Indeed, many historians believe that Lafayette eventually became involved in the American Revolution largely because his father had died in one of those wars between the French and the British.

Thus, when British subjects in North America fomented a rebellion against their mother country, the French followed the event with great interest. In addition, Enlightenment writers had produced works on liberty that had profoundly influenced a great many French citizens. These individuals understood the American Revolution as the logical culmination of these writings, and as a result they tried to keep abreast of developments in North America. For these reasons, Lafayette would have had access to plentiful information about the American Revolution.

3. What was the nature of his involvement in the American Revolution?

It seems that the summer of 1775 set Lafayette on his course to aiding the American cause. At that time, the French Army went to train in the town of Metz, and at a dinner one night he and his commanding officer talked about the rebellion that had recently broken out in British North America. That year also saw him become a Freemason, and the principles of that organization also resonated with the goals that the Americans sought. Interestingly, there were quite a few prominent American revolutionaries who were Freemasons, among them, George Washington, Benjamin Franklin, and Paul Revere.

By December of 1775, Lafayette said that his "heart was dedicated" to the American cause. For that reason, Lafayette may have eventually decided to go North America on his own (much as Kosciuszko did). As it turned out, he would actually receive the blessing of King Louis XVI to embark on that course of action. Eager to weaken his traditional enemy, the French monarch surreptitiously began to encourage officers in his army to offer their services to the rebelling colonies. When he became aware of this, Lafayette used his influence to become one of those individuals. Accordingly, in December 1776, the American representative in France offered him a commission in the Continental Army as a major general. He accepted and joined—at the age of 19!

4. In what capacity did he serve and for how long?

Lafayette would indeed serve in the American Army as a major general, but political developments in France almost prevented him from ever reaching North America. When the British found out that the king had encouraged French officers to aid the Americans, they told the French ruler that such an action would constitute an act of war. Not willing at that point in time to engage in such a conflict, the king rescinded his backing for Lafayette and others to go to America.

Lafayette acquiesced for four months, but in April of 1777 he decided to disobey his orders. Purchasing his own ship, he sailed to America, landing in South Carolina. He then journeyed to Philadelphia, and presented his appointment as a major general to Congress. Initially, they rebuffed him, stating that they did not have the money to pay him. Lafayette offered to serve without pay, and further ingratiated himself by demonstrating a knowledge of the English language that few foreign volunteers possessed. Congress then honored his commission, confirming the rank of major general.

Soon, George Washington arrived in Philadelphia to speak to Congress, and at a dinner he met Lafayette. Immediately impressed with the young Frenchman, Washington soon made him a part of his staff. A month later, Lafayette saw action at the Battle of Brandywine (September 1777), suffering a leg wound while stabilizing the American flank. Once Lafayette recovered from his wound, Washington attached him to a force commanded by Major General Nathaniel Greene. In this capacity, Lafayette won a victory against Hessian forces at Gloucester, New Jersey, in November 1777.

He spent the brutal winter of 1777–1778 with Washington and his troops at Valley Forge. He used his own money to buy uniforms and muskets for his troops and he shared the hardships of the weather with his troops. Although his young wife encouraged him to return to France, Lafayette remained committed to his soldiers, General Washington, and the colonial cause.

During the next three years, Lafayette served primarily in a subordinate role to other leaders, but in early 1781 Washington gave him command of a force tasked with stymieing the activities of the British general Lord Cornwallis in Virginia. Eventually, his efforts helped Washington pen up Cornwallis at Yorktown. The surrender of Cornwallis there in October of 1781 would cause the British to give up their efforts to subdue the Americans, leading to our independence.

5. Obviously when the war ended, he returned to France. What did he do at that time?

After the British surrender at Yorktown, Lafayette returned to France to a hero's welcome. As the war had not yet ended, he received orders from King Louis XVI to begin planning an assault on British possessions in the West Indies. However, before this could take place, on September 3, 1783, the Americans and the British signed a peace treaty, the Treaty of Paris, ending the war. Lafayette then returned to civilian life, but four years later he would become embroiled in the growing turmoil within his own country. Needing money, King Louis XVI called in 1787 for a meeting of highly-placed Frenchmen to help him raise funds necessary for him to run the country.

Because of his noble status, Lafayette served as a delegate. While serving in this capacity, Lafayette called upon the king to reform the French political system. Nothing came of this effort on his part. Because the delegates had not proven able to help the king, in 1789 Louis XVI called the French national assembly (known as the Estates-General) into session. Once again, Lafayette served as a delegate. Rather than raise taxes, many of the delegates chose instead to demand more representation.

Despite his aristocratic title, Lafayette sided with the commoners. Eventually, this movement led to the French Revolution. Lafayette sided with the rebels, and offered them a document that he drafted and titled "The Declaration of the Rights of Man and of the Citizen." In gratitude, the leaders of the rebellion made him commander of the national army. Unfortunately, Lafayette found himself trapped between his loyalty to the king and his belief in the principles of the French Revolution.

He gradually fell out of favor, and eventually wound up as a political prisoner. When he received his release from captivity, Lafayette avoided any further political involvements until 1814. Elected to France's Chamber of Deputies in 1815, he helped to arrange the process by which Napoleon Bonaparte agreed to go into exile. For the next twenty years, Lafayette tried to help his country become both more stable and democratic. Unfortunately, it seemed that he had accomplished neither task when he died in 1834.

6. We obviously owe him a great deal. How has the U.S. honored this gentleman?

Without question, Lafayette is the most revered of all the foreigners who aided the American cause. The best proof of this stems from the fact that upon Lafayette's death, American President Andrew Jackson gave him the same memorial procedures that had been followed when George Washington died. Since then, Americans have continued to honor him. Five states have cities named for him, and numerous parks and streets bear his name. He has graced the front of a number of United States Postal Services stamps, and three ships named the U.S.S. *Lafayette* proudly served in the U.S. Navy.

7. Have books been written about him? Where should one start to read about him?

Quite a few books have examined the life of Lafayette. James Gaines wrote *For Liberty and Glory: Washington, Lafayette, and Their Revolutions* in 2007, and most historians generally regard this as the best work on Lafayette.

8. What have I neglected to ask?

When the first American forces arrived in France during the First World War to assist in the struggle against Imperial Germany, an American officer, Colonel Charles E. Stanton, stood at Lafayette's tomb in Picpus Cemetery in Paris and gave a speech in which he said “Lafayette, nous voilà” meaning, “Lafayette, we are here.” Widely reported and often referenced more than one hundred years later, these words are a very fitting tribute to a person who selflessly aided our country’s efforts to earn its freedom.

CHARLES LEE

G. R. Hall engraving, 19th century
Image, public domain

TWELVE

Charles Lee

TREACHEROUS GENERAL ★ (1732–1782)

1. Born in England, Charles Lee lived in North America at the start of the American Revolution. When and where exactly was he born, and how did he come to the colonies?

In a manner similar to that of a number of George Washington's other generals, Charles Lee was born in Great Britain—specifically, in the English community of Dernhall, Cheshire, England on February 6, 1732. Lee's father held a high rank in the British Army, and his family's affluence allowed Lee to attend school in Switzerland. While there, he learned a number of languages. When he turned 14, Lee came back to England and enrolled in the King Edward VI School in Suffolk. He did not remain there long, however, as a year later his father paid to have him commissioned as an ensign in the British Army (the purchase of commissions was standard practice in the British Army until the 1871). Not surprisingly, Lee became a member of his father's regiment—the 55th Regiment of Foot.

Four years later, his father died, and Lee then used his own money to purchase the higher rank of lieutenant for himself. In 1754, his regiment received orders to sail to North America after news of the outbreak of hostilities between the French and Virginians reached Great Britain in what was known in the colonies as the French and Indian War, and in Europe, as the Seven Years' War. Lee and his regiment became a part of Major General Edward Braddock's campaign to capture Fort Duquesne, which the French had constructed on the site of present-day Pittsburgh. Interestingly, other participants in that campaign included future American generals George Washington, Daniel Morgan, and Horatio Gates.

After Braddock's bloody repulse, the British high command transferred the remnants of Lee's regiment to upstate New York to coordinate with Native American tribes allied with the British. Soon thereafter, Lee bought himself a commission as a captain. After his stay in the Mohawk Valley, in 1757, Lee participated in a failed British campaign to capture the French fortress of Louisburg on Cape Breton Island. A year later, Lee and his regiment took part in an assault on Fort Carillon. Lee suffered a wound, and the attack proved unsuccessful.

After recuperating for nearly a year, Lee returned to duty in time to participate in the capture of Fort Ticonderoga, and in 1760 he and his regiment helped force the surrender of the French garrison defending Montreal. While that event essentially ended combat between the French and the British in North America, fighting continued in Europe, and Lee received orders to report for duty in Portugal. There he fought bravely, especially earning praise for his conduct at the October 1762 Battle of Vila Velha. In another interesting coincidence, John Burgoyne, who would surrender a combined British-Hessian army at Saratoga in October of 1777, served as Lee's commander at Vila Velha.

A major when the conflict ended in 1763, Lee soon realized that he had little chance of receiving another promotion as things stood within the British military. To improve his chances for advancement, he fought briefly with the Portuguese army against the Spanish invasion of Portugal (1762). Lee went to Poland to serve as a military advisor to the Polish king Stanislaus II in 1765. Two years later, he returned to England, but found his prospects for promotion no brighter. When Poland became embroiled in the Russo-Turkish War (1768–1774), Lee made his way back to that country, and served as a major general. This earned him a promotion to the rank of lieutenant colonel in the British Army, but despairing of ever receiving another promotion, Lee decided to resign his commission and in 1773, he moved to North America.

2. How did he first get involved in the American Revolution?

When Lee arrived in what is now West Virginia, he found many of his fellow settlers very dissatisfied with their relationship to the British government. Interested in both gauging the depth of this anger and acquainting colonial leaders with his agreement with their grievances, Lee traveled throughout the colonies for the next few months. This effort by Lee paid immediate dividends after the Battles of Lexington and Concord. When the Second Continental Congress convened in May of 1775, it quickly took up the subject of who to appoint as the commander of Patriot forces, and Lee's name immediately surfaced.

A number of the representatives thought Lee had the requirements to receive command, but felt that the honor should go to someone born in the colonies. In addition, Lee's personality proved repugnant to a number of the representatives. As a result, George Washington became the commander, and Congress appointed Lee as a major general, third-in-command of the American Army. When Artemis Ward retired in early 1776, Lee became Washington's second-in-command.

3. What were some of Lee's victories and exploits, and on the other hand, some of his defeats and disasters?

Upon receiving his initial appointment from Congress, Lee journeyed to Boston to join the army. While stationed there, Lee received notification from Congress that it had appointed him to command the American Army's Canadian Department. Whether because of the deteriorating situation there or a more pressing need elsewhere, Congress rescinded its original order, choosing instead to give him command of the Southern Department. When Lee arrived in Charleston, South Carolina in early June of 1776, he learned of an impending British operation to capture the city. In the next few weeks, under his direction fortifications sprang up around the harbor, including an installation that became known as Fort Moultrie.

Fearing that the soft wood and sand that comprised its outer walls could not withstand a bombardment from British naval guns, Lee recommended that the defenders abandon the fort. Civilian authorities countermanded this request—fortuitously for the Americans, as the fort played the most prominent role in repelling the British assault on June 28, 1776. Although his strategic vision almost doomed the American cause at Charleston, he received great acclaim as the defender of the city.

Ordered to rejoin Washington's army in the north, Lee arrived after the British had routed the Americans at the August 27, 1776 Battle of Long Island. Apparently, this further convinced Lee of his superiority over Washington as a military leader. Any doubt of this that Lee had left was ended when Washington chose to keep 3,000 troops in a fortification (ironically, named Fort Lee in his honor) on the Hudson River that the British captured with little difficulty. As Washington retreated out of New York City and through New Jersey, Lee kept forces under his direct command at a distance from the main body of the American Army.

Lee later defended his actions on a tactical basis, but most historians believe that Lee chose to remain separate from Washington to show his personal disdain for his commander. Situating his force near Morristown, New Jersey, Lee then decided to move his headquarters and promiscuous social activities to a nearby tavern. There, on the morning of December 15, 1776, a well-coordinated British cavalry raid led by Banastre Tarelton and 16th Queen's Light Dragoon captured Lee. He would remain a prisoner of the British until exchanged for British Major General Richard Prescott in April 1778.

4. As we all know, personality differences often occur and contribute to one's downfall. What do we know about this individual's personality and his clashes with General Washington?

Ironically, when the British captured Lee, they interrupted him in the process of writing a letter to Horatio Gates about Washington's failures as a general. Since the start of the American Revolution Lee had believed that he outclassed Washington as a military leader in every respect. One could argue, as Lee's defenders have done, that he had seen none of the tactical and strategic brilliance that Washington would display in the Trenton-Princeton campaign before his capture in mid-December 1776.

On the other hand, Lee had seen Washington's superb leadership qualities in the aftermath of Braddock's defeat, when the young Virginian had ably mounted a rear-guard action to save the remnants of the British force. Lee's haughtiness therefore seems based more on his massive ego than on the actual evidence at hand. Surprisingly, even though Washington learned of Lee's condescension towards him, Washington tried to remain on cordial terms with his subordinate. As we shall see, Lee's actions in 1778 finally made it impossible for Washington to continue to tolerate his second-in-command's insubordination.

5. What is the old saying "Old soldiers never die—they just fade away"? How does the famous saying "Old soldiers never die—they just fade away" relate to General Lee?

After his parole granted by the British, Lee returned to Washington and reported for duty in May of 1778. From all accounts, he looked at the ragged condition of the soldiers at Valley Forge as further proof of Washington's incompetence as a military leader. A month later, as the British began preparations to evacuate Philadelphia, Lee learned that Washington planned to attack the British column as it moved through New Jersey on its way to New York City.

According to military protocol, Washington gave direct command of the assault to Lee as the second-in-command. Having only seen Washington's army as it retreated after losing the Battle of Long Island and in its disheveled state at Valley Forge, Lee immediately found fault with Washington's plan, and declined to lead the attack. Unfazed by his refusal, Washington then gave command of the operation to the French ally and general Gilbert du Motier, Marquis de Lafayette.

Considering this a slight, Lee then changed his mind, and agreed to lead the attack, and on the morning of June 28, 1778, he led a force of 4,500 soldiers forward into battle at Monmouth Court House. Initially pleased with the progress his troops made, Lee quickly saw his operation begin to fall apart. Either because he honestly thought the British had legitimately gained the upper hand on the battlefield or because of his low estimation of the soldiers he commanded, Lee then ordered his men to retreat. Ready to move the bulk of the army into battle as Lee's attack disrupted the British retreat, Washington became perplexed when he heard the sound of battle getting closer to, rather than further from, his position. Riding forward, Washington found his army retreating.

Confronting Lee, Washington asked for an explanation of the withdrawal. Lee offered excuses, but Washington, displaying a temper rarely seen during the war, furiously ordered Lee to the rear and took personal command of all the troops on the battlefield. Displaying the superb training that Baron von Steuben had given them at Valley Forge, the American soldiers stopped their retreat, reformed, and held their positions for the rest of the day.

Most generals in Lee's position would have simply been thankful that their army had averted a catastrophe, and would lay low until the anger over their actions had subsided. Instead of adopting that sensible strategy, Lee chose to write a letter to Washington, stating that he had saved the American Army from a disaster by retreating. Moreover, he insisted that Washington needed to give him an apology for maligning his conduct during the battle. Having reached the limits of his patience, Washington informed Lee that he would request a formal inquiry into Lee's conduct. That caused Lee to demand a court martial, which found Lee guilty on three counts, and suspended him from the army for a year.

Lee's criticism of Washington and Lee's sharp tongue was not well received by fellow officers. Seeking to defend Washington's honor, in 1778, Colonel John Laurens fought a duel with Lee on the edge of a wood near Philadelphia on December 23. In the duel, Lee was slightly wounded in the side, after which the two officers stopped the duel.

Though Lee suffered a non-life threatening wound, something that would have convinced many individuals to stifle their criticisms, he was not stopped in his animosity toward Washington. Lee, who had lost two fingers in a duel in Poland years earlier, chose instead to raise the stakes, taking his argument to Congress. Finding his letters insulting, Congress removed him from the American Army in 1780, thus ending his military career.

6. What were his later years like, and when and where was his buried?

After his dismissal from military service, Lee went back to live on his acreage in present-day West Virginia. In 1782, he made a journey to Philadelphia. While there, he developed a high fever, and died on October 2, 1782. He was buried in Philadelphia. Lee had lived long enough to realize that the British surrender at Yorktown the previous

October had all but guaranteed American independence, but not long enough to actually see the Treaty of Paris signed on September 3, 1783.

7. What else can be said about this general, who seemed to be insubordinate at times?

Although he died at the age of 50, Charles Lee managed to lead a life filled with many fascinating twists and turns. Brave enough to distinguish himself in battle and face death in two duels, Lee also possessed an arrogant self-confidence that seems totally misplaced. Furthermore, questions have emerged about his loyalty to the American cause. In the 1850s a librarian discovered a letter that seems to suggest that while he was a British prisoner he offered them advice on how to crush the American cause.

Moreover, some historians believe that the ease with which the British captured him in December 1776 suggests that he himself may have tipped the British off to his location, giving himself an "honorable" exit from a cause seemingly on its last legs. In an interesting twist, some historians have suggested that Lee's enemies inside the American Army may have supplied that bit of intelligence to the British to get rid of him. Lastly, few would know that Charles Lee had a distinction that no other of Washington's generals could claim. While serving during the French and Indian War in upstate New York, Lee had immersed himself in the ways of the Mohawk Tribe, to the point that they gave him the name "Boiling Water." And in a manner virtually unheard of at the time, Lee married the daughter of a Mohawk chief.

Apparently, a woman of great beauty, she bore him twins. Sadly, little outside of comments by Lee and others on her attractive features and an account of the birth of his children, no evidence exists of what happened to either his wife or his children and their names are unknown. One can only hope that someone will eventually uncover more information about these people lost to history!

HENRY "LIGHT-HORSE HARRY" LEE III

By Charles Wilson Peale, 1782
Independence Hall, Philadelphia (INDE 14041)
Image, public domain

THIRTEEN

Henry "Light-Horse Harry" Lee III

TENACIOUS CAVALRY COMMANDER ★ (1756–1818)

---

1. We sometimes have very iconic, interesting names for our heroes. In the American Revolution, we had the "Swamp Fox," and later in the Civil War there was "Stonewall" Jackson. Now we learn of "Light-Horse Harry" Lee. Where was he born and how did he get involved in the American Revolution?

Henry Lee III was born on Leesylvania Plantation near Dumfries in Prince William County, Virginia, on January 29, 1756. His family traced its Virginia lineage back to the 1600s, and both Henry Lee I and Henry Lee II had served as officers in the Virginia militia. The Lee family's wealth allowed them to send Henry to the College of New Jersey, what is now Princeton University. He graduated from there in 1773 at the age of 17, and chose to pursue a career in the legal profession. His days as a lawyer did not last long, however, as he immediately offered his services to the Patriot cause upon the outbreak of the American Revolution in 1775.

2. How did his famous nickname come about for him?

When he made the decision to join the American cause, Lee enlisted in a Virginia cavalry regiment in June 1776. Given the rank of captain in the Virginia Light Horse, Lee commanded the 5th Troop in that regiment. Soon, Washington assimilated the unit into the regular US Army, and renamed the regiment the 1st Continental Light Dragoons (also known as Bland's Horse because the commanding officer was Lee's cousin, Colonel Theodorick Bland). At that point in time, armies usually featured two types of mounted soldiers. Members of cavalry units attacked on horseback and usually used swords to assault their enemies.

Dragoons, on the other hand, rode horses into battle but usually dismounted and fought on foot. Given this role, Lee and his men soon distinguished themselves with their ability to ride quickly into battle and engage the enemy. Because of that ability, the army soon nicknamed him "Light Horse Harry." Throughout the rest of the war, his exploits would guarantee that the sobriquet remained with him.

3. Apparently he had his own squadron of men—Lee's Legion. Tell us about them and their heroics.

Because of Lee's distinguished service with the 1st Continental Light Dragoons, Washington decided in 1778 to create a roving reconnaissance unit and recommended to Congress that Lee command it. Congress agreed and promoted then Major Lee to Lieutenant Colonel Lee and give him an independent command. This consisted of the 100 cavalry troops belonging to his 5th Troop, along with 180 infantrymen. Known as Lee's Legion, this unit saw action for the first time in September 1778 when Lee's men attacked a detachment of Hessian soldiers in Hastings-on-Hudson, New York. Lee's Legion inflicted 23 casualties, while suffering none themselves.

A year later, on August 19, 1779 Lee's Legion attacked a British fort at Paulus Hook, New Jersey, and succeeded in capturing 158 British soldiers. Impressed with his accomplishment, Congress awarded him a gold medal. While Congress gave out a number of such medals,

Lee was the only officer below the rank of general to receive such an award. These actions made Lee's Legion one of the most famous units in the Continental Army during the Revolutionary War.

4. At one point during the war Lee fought with General Francis Marion, the legendary "Swamp Fox." What does history tell us about those days?

In May 1780, the British had succeeded in capturing Charleston, South Carolina. They then moved inland, and on August 16, 1780 inflicted a crushing defeat on an American force at the Battle of Camden. Unwilling to simply cede the South to the British, George Washington sent military units to that theater of operations to try to stabilize the situation. Determining that this force needed a cavalry contingent, Washington ordered Lee to take his green-coated legion to South Carolina.

When he arrived in South Carolina, Lee immediately sought a rendezvous with Francis Marion. A lieutenant colonel in a South Carolina regiment, Marion had managed to avoid capture at Charleston and Camden, and had afterwards engaged in a campaign of guerilla warfare against the British. Nicknamed the "Swamp Fox" for his ability to elude British attempts to capture him, Marion conducted raids on British outposts in the hinterland of South Carolina. When Lee met up with Marion (the latter by then promoted to brigadier general), the two agreed to launch an attack on one such British position at Georgetown, South Carolina in January 1781.

Nothing came of this campaign, but later that year the two did succeed in capturing Fort Watson and Fort Motte respectively in April and May 1781. These actions wreaked havoc on British communications throughout the state and forced them to adopt a new strategy for suppressing American resistance. In this manner, two of the most marvelously nicknamed figures in American history lent their combined talents to efforts that would help to resurrect their nation's cause in the South. Lee was present at the surrender of British forces by Lord Cornwallis at Yorktown on October 19, 1781.

5. After the Revolutionary War ended, what events continued to engage him?

Lee actually left the army before the end of the war, and returned to the practice of law. Soon, however, his fellow citizens elected him to represent the state of Virginia as a delegate of the Continental Congress from 1786 until 1788. In 1789, he became a member of the Virginia state legislature, serving in that body until 1791. He then won his state's gubernatorial race, and would serve as governor for three years. His political career went on hiatus in 1794 when George Washington (by then the president of the United States) asked him to command U.S. military forces tasked with suppressing what is known in American history as the Whiskey Rebellion in western Pennsylvania.

Four years later, President John Adams, fearing a possible war with the French, commissioned him as a major general in the regular U.S. Army. Nothing came of this, however, and Lee returned to civilian life. Shortly thereafter, Lee won election to the U.S. House of Representatives. He served one term, and then returned to civilian life. Unfortunately, his life soon took a turn for the worse. His business ventures proved unsuccessful, and a mob beat him in 1812 because of his opposition to the war (War of 1812) that had recently broken out with the British. Lee never fully recovered and died a few years later on March 25, 1818.

6. Where is he buried? I understand he received full military honors!

In an effort to regain his health, Lee had gone to the Bahamas after the War of 1812. The trip did not help, however, and Lee decided to return to Virginia. But on his return trip, his health deteriorated, and

Lee disembarked at a plantation owned by the family of Revolutionary War General Nathaniel Greene. In March of 1818, he passed away there. Learning of Lee's death, President James Monroe ordered nearby U.S. Navy forces to sail there and give Lee a burial with full military honors. In 1913, it was decided that his remains should be disinterred and moved to a chapel on the campus of Washington and Lee University. There he was buried beside his second wife and his much more famous youngest son: Confederate General Robert E. Lee.

7. What have I neglected to ask about this famous American hero?

Henry Lee III is an example of the fleeting nature of fame in American history. Once a household name for his courageous service during the American Revolution, he became essentially a footnote as time passed. If they remember him at all, most Americans know him only as the father of Robert E. Lee (1807–1870).

In addition, they may know the toast, originally given not as a toast but as a funeral oration to Congress at Washington's death: "To George Washington: first in war, first in peace, and first in the hearts of his countrymen," but few will know that Lee first said those words. Americans should do more to commemorate people like him who helped to give us a nation during the first critical years of our existence.

BENJAMIN LINCOLN

By Charles Wilson Peale, 1781–1783
Independence Hall, Philadelphia (INDE 14097)
Image, public domain

FOURTEEN

Benjamin Lincoln

LOYAL SUBORDINATE ★ (1733–1810)

1. The name Abraham Lincoln certainly stands out in American history. But Abraham Lincoln had an unrelated predecessor, Benjamin Lincoln, who had a major part in the American Revolution. What do we know of his birth, heritage, and early years?

Although Abraham Lincoln appears to have not known much of his family's history, genealogists proved able to trace his ancestry back to Samuel Lincoln, who arrived in Massachusetts in 1637. Interestingly, at the same time that Samuel Lincoln arrived in the Massachusetts community of Hingham, another family with that same last name also settled there. Thorough investigations have determined that this second family of Lincolns bore no relation to Samuel Lincoln. The latter set of Lincolns proved remarkably successful in Hingham, becoming one of the most financially prominent families in that community.

Indeed, because of the family's stature, Benjamin Lincoln's father (also named Benjamin) served on the Massachusetts Governor's Council and held the rank of colonel in the colonial militia. Thus, when Benjamin Lincoln was born on January 24, 1733, he entered into a life of wealth and privilege. His family's wealth allowed him to attend the local school, but although his family could have afforded it, Lincoln chose not to attend college.

2. What were some of his early military experiences like?

When hostilities between the French and the British escalated in the Ohio Territory in 1755, Benjamin Lincoln chose to become a part of the Massachusetts militia. Because his father commanded the 3rd Regiment, Benjamin Lincoln joined that unit, becoming his father's adjutant. Although he personally did not participate in any military engagements during the French and Indian War, he performed his other duties well, as he received promotions to the rank of captain and then major.

After hostilities ceased in 1763, Benjamin Lincoln remained active in the militia, and in 1772 he became the lieutenant colonel of his regiment. His military service seemed to have placed him in a position where he could participate as a combatant when Patriot resistance began in Massachusetts with the Battles of Lexington and Concord, but as it turned out Lincoln's aid to that cause came first in the political realm.

Starting with his election as a selectman in Hingham in 1765, Lincoln had embarked on a career of holding successively higher offices in his community. This process culminated with his election to the Massachusetts Provincial Congress in 1774. Placed on committees that oversaw the militia in general and their supplies in particular, Lincoln proved able to help fill the needs of the force besieging the British in Boston after Lexington and Concord. Based on this service, Massachusetts gave him an appointment as a major general of militia in January of 1776. Given the responsibility of defending the coast of Massachusetts, Lincoln helped drive away the last British ships from Boston Harbor in May of that year.

Having accomplished that task, Lincoln sought a commission in the Continental Army. While he waited for the appointment, Lincoln took a brigade of Massachusetts militiamen to aid George Washington in his defense of New York City. Arriving after Washington's defeat at the Battle of Long Island in August, Lincoln's men joined the American force as it began its retreat from New York to Pennsylvania. Even though he had seen no combat while serving under Washington, he had made a favorable impression on the commanding general, who recommended that Congress give him a commission. In February of 1777, Congress did so, appointing him as a major general. Washington then gave Lincoln command of a detachment guarding Bound Brook, New Jersey.

His isolated position at Bound Brook proved a tempting target for the British, who attacked Lincoln there in April. Caught completely by surprise, Lincoln and his men beat a hasty retreat, with Lincoln leaving his personal papers and his baggage. Despite this defeat, Washington seems to have assigned no blame to Lincoln for the mishap, as he soon gave Lincoln an assignment of vital importance in New York.

3. Benjamin Lincoln played an important role in campaigns that resulted in the capitulation of British forces at Saratoga, Charleston, and Yorktown. Why were these events important?

When British General John Burgoyne led an army out of Canada with the goal of gaining control of the entire Hudson River valley, George Washington had responded by sending what troops he could spare to join the American force already in New York. Upon his arrival there, Lincoln received orders to use volunteer militia units from New England to prey upon Burgoyne's lengthening supply line. Lincoln did so, until receiving orders from General Horatio Gates to join him north of Saratoga. Although he saw no action at the Battles of Freeman's Farm in September 1777 or Bemis Heights in October 1777, Lincoln did deploy his troops under his command in a manner that effectively supported the portions of the army that did engage the British.

When it became apparent that Burgoyne would attempt to retreat to Canada, Lincoln proposed that the force under his command block a ford across the Hudson River. This prevented Burgoyne from effecting an escape, and forced him to surrender his army to Gates in October 1777.

In 1778, Washington gave Lincoln command of American forces in the Southern Department which was a large and independent command in the southern colonies.. When French forces arrived to assist Patriot forces in 1779, Lincoln moved south to siege and try to recapture Savannah, Georgia. A lack of coordination between the French and Americans resulted in a repulse by the British defenders, and Lincoln then returned to Charleston, South Carolina. The following year,1780, a British force launched a campaign to capture that city.

Earlier, in 1776, American forces had successfully defended Charleston, but in 1780 the British proved able to besiege the city. Seeing no other viable course of action, Lincoln surrendered the city and all its defenders (more than 5,000) in May. It was the largest surrender of American troops during the American Revolution. A court of inquiry found that Lincoln had done all that he could to defend the city, and therefore recommended that his face no disciplinary action for the loss. Held by the British as a prisoner of war until exchanged for an officer of equal rank captured at Saratoga, Lincoln rejoined the American Army in November 1780.

Washington welcomed Lincoln back, and made him his second-in-command for the Yorktown Campaign. When Washington moved south in the fall of 1781 to trap the British at Yorktown, Virginia, Lincoln ably kept his soldiers disciplined on the line of march. Commanding a force augmented by thousands of French soldiers and aided by the timely arrival of a French fleet, Washington besieged a British army under the command of General Charles Cornwallis.

Recognizing the futility of further resistance, the British surrendered on October 19. Mortified by this humiliating defeat, Cornwallis refused to participate in the formal surrender ceremony, sending his second-in-command instead.

When he realized this, Washington refused to accept the surrender, telling the British officer to surrender to his second-in command—Benjamin Lincoln. Lincoln thus found himself present at the surrender of three armies during the American Revolution. The British surrenders had momentous consequences. Indeed, Burgoyne's capitulation had convinced the French to aid the American cause, while the surrender at Yorktown ultimately proved decisive, as it eroded British civil support for the military campaign to subdue the Americans.

4. Certainly many soldiers, including a number of Washington's generals, were wounded during the American Revolution. But Benjamin Lincoln seemed to have suffered a particularly grievous wound. What happened?

As we have seen, Benjamin Lincoln did not see combat until the Battle of Bound Brook in April 1777, but he fled the battlefield before British soldiers could close in on him. His next encounter with the British came after the Battle of Bemis Heights, when he moved his force to block the ford across the Hudson while helping fortify Fort Edward.

In the darkness, he and his men ran into a British force, and both sides opened fire. One British musket ball hit his right ankle, shattering the bone. By the time his wound healed after a painful recovery, Lincoln's right leg was two inches shorter than his leg—the same thing that had happened to Benedict Arnold after his left leg wound suffered a few days before at the Battle of Bemis Heights.

5. Did Benjamin Lincoln play any important roles after the American Revolution?

Even before the end of the American Revolution, Benjamin Lincoln had started a transition back into public life. After the Second Continental Congress adopted the Declaration of Independence, it began deliberations on a blueprint for the government that would represent the United States. This process resulted in the creation of The Articles of Confederation and Perpetual Union. One provision of this document established a Department of War, and after The Articles of Confederation and Perpetual Union went into effect in 1781 Congress appointed Benjamin Lincoln as the first Secretary of War.

In 1783, Lincoln left that position and returned to Massachusetts. Four years later, Lincoln found himself once again tasked with a military responsibility, but on this occasion he led troops against his fellow countrymen. This came about because of an incident known as Shay's Rebellion. Due to demands for payment of debts in hard cash by creditors in Massachusetts after the American Revolution instead of the paper money that circulated during the conflict, a number of farmers in the western part of the state refused to pay their creditors. This so alarmed the state's wealthy residents that they raised money to field an armed force to suppress the rebellion. These financiers then prevailed upon Lincoln to help lead a force of 3,000 against those in rebellion. He did so and was successful in this endeavor.

For many individuals, involvement in an action against American citizens may have cost them political capital, but apparently not in Lincoln's case. Proof of this lies in the fact that after Shay's Rebellion he ran for the office of lieutenant governor, and won. After leaving office, he became the Collector for the Port of Boston. He served in that capacity until 1809, when he decided to return to private life.

6. Where is he buried, and how is he currently acknowledged?

Unlike most of Washington's generals, Benjamin Lincoln lived to a very old age by the standards of the times, finally passing away in May of 1810 at the age of 77. He was buried in the cemetery located at the Old Shipyard Church in Hingham, Massachusetts. At the time an unremarkable structure, the Old Shipyard Church today has the distinction of serving as the oldest Puritan meetinghouse still standing. Incredibly, Lincoln's house in Hingham still remains as well. After his death, a number of communities, primarily in the South, named themselves after him. Finally, befitting the role that he played in the battle there, the U.S. Coast Guard Training Center at Yorktown has a Lincoln Hall.

7. What have I neglected to ask about this famous general?

A popular movie of the late 1990s was titled *Almost Famous.* This seems a fitting term to apply to Benjamin Lincoln. On paper, he seems to have played a significant role in the two surrenders of British forces during the American Revolution, as he served as the second-in-command on both occasions.

He also had the unfortunate distinction of having to surrender the largest number of U.S. soldiers in our nation's history until the mass capitulation on the Bataan Peninsula in 1942. But perhaps because he played supervisory roles throughout the war, and largely failed when given command of forces in action, Lincoln has come down through history as merely one who was useful rather than crucial to the eventual American triumph.

FRANICIS MARION

Engraving, 1880
New York Public Library (#430597)
Image, public domain

FIFTEEN

Francis "Swamp Fox" Marion

FATHER OF GUERILLA WARFARE ★ (1732–1795)

1. No discussion of the American Revolution would be complete without including the "Swamp Fox," Francis Marion. What do we know about his early life?

Like many of the other American generals from the Revolutionary War that we have discussed, Francis Marion came to the Patriot cause from an affluent background. His family owned a plantation in Berkeley County, South Carolina, and in 1773, Marion purchased his own plantation (which he named Pond Bluff). Unlike the other generals we have studied, however, we do not know the date—or even the year—of his birth. Most historians assume that he was born in 1732. Along similar lines, we know that Marion had a leg disfigurement, but we do not know specifically what had caused it. We thus know a great many things about Marion, with some notable exceptions.

2. Apparently, he went to sea as a youth, but later decided to stay on dry land. What happened?

By all accounts, as a youth Marion had a bit of wanderlust to him. Accordingly, he chose at the age of 15 to join the crew of a ship hauling cargo to the West Indies. Unfortunately, the ship sank on his first voyage, supposedly from getting rammed by a whale. Marion and the rest of the crew managed to escape in a lifeboat, but spent seven days at sea before a passing ship rescued them. This incident seemed to have ended Marion's infatuation with a nautical life, as he never put to sea again.

3. What was the extent of his involvement in the French and Indian War? And how did his exploits there prepare him for later endeavors?

Along with many of the other generals that we have studied, Marion had his first military experience during the French and Indian War. As previously noted, this conflict had its origins in a foray into the Ohio Territory led by Virginia Militia Colonel George Washington. Most Native American tribes in North America wound up fighting on the side of the French in this conflict, including the Cherokees. Soon, that tribe began to attack settlements in the South Carolina hinterland. In response, the colony sought to recruit new members for its militia, and in 1757 Marion volunteered for service.

Sent out to the frontier with his unit, Marion engaged in brutal irregular warfare against the Cherokees until they sued for peace in 1761. He would later apply much of what he had learned while fighting the Cherokees to the prosecution of his guerilla campaign against the British during the Revolutionary War.

4. What were his initial forays into battle like during the American Revolution?

When South Carolina learned of the Battles of Lexington and Concord, the colony responded immediately by authorizing the formation of three regiments of militia for use against the British. Because of his military service against the Cherokees, Marion received a captaincy in the 2nd South Carolina Regiment. Recognizing that the British might at some point attack Charleston, South Carolina, the regiments began to fortify the port city.

As part of this effort, Marion and his regiment constructed Fort Sullivan in the harbor of Charleston. This precaution proved fortuitous, as a British fleet did indeed attack Charleston in June 1776. Perhaps underestimating their enemy, the British chose to sail directly into Charleston Harbor and attempt to shell the American fortresses into submission. But the South Carolinians withstood the bombardment, and after a spirited battle, repelled the British fleet.

According to all accounts, Marion fought bravely in this engagement. Three years passed before Marion once again saw combat. In 1779, a combined Franco-American force that included Marion's regiment attempted to capture Savannah, Georgia, but the effort failed. He would not see action again until the summer of 1780, when he began a guerilla campaign against the British that would earn him his famous nickname.

5. Speaking of which, how did he get the name the "Swamp Fox"?

In 1780, the British once again attempted to capture Charleston. This time, however, they landed an army on the coast that quickly encircled the city. At the same time, the British navy effectively shut off all contact with the city by sea. After a prolonged siege, the American force defending Charleston surrendered in May 1780. Dismayed by this development, the Continental Congress dispatched a force under the command of Horatio Gates to retake the city, but he suffered a disastrous defeat at the Battle of Camden. These two setbacks meant that for all intents and purposes the Patriot cause had no organized military presence in South Carolina, as the British had killed or captured virtually every American soldier.

Fortunately, Marion had avoided either of these fates. After breaking his leg in the spring of 1780, Marion had returned to his plantation to recuperate. By the time he did, Charleston had surrendered, and Gates had lost. Rather than despairing, Marion chose instead to raise a company of men to fight an irregular war against the British. He and his men would hit isolated British units, and then seemingly vanish into the swamps.

A British officer named Banastre Tarleton tasked with capturing Marion once pursued him for twenty-five miles until suddenly finding that his trail had gone completely cold. Astonished, the British officer told his men that "as for this damned old fox, the Devil himself could not catch him." In this manner, Marion earned one of American history's most famous nicknames.

6. Some believe that the movie *The Patriot* was somewhat based on his life. Are there any resemblances or overlap?

Clearly, Mel Gibson based much of the movie *The Patriot* on the exploits of Francis Marion. Like Marion, Gibson's character had fought in the French and Indian War, and engaged in a guerilla campaign against the British in South Carolina. Hollywood and history differ in a number of ways, however.

For example, in the movie Gibson's character comes late to the Patriot cause, while as previously noted Marion joined the Patriot cause in 1775. Gibson's character becomes a Patriot when the British kill one of his sons, but Marion had no children at the time of the Revolutionary War.

Finally, to the chagrin of many historians, Gibson's character owns no slaves, while Marion did. But the movie does do an excellent job of showing how a handful of Americans led by Marion helped keep the cause of liberty alive in the face of daunting odds in South Carolina in 1780.

7. Banastre Tarleton (1754–1833) was allegedly sent by King George III to deal with the Swamp Fox. Did he ever succeed?

Banastre Tarleton had arrived in North America in 1775 as a lieutenant in a British cavalry regiment, and by 1778 he had risen to the rank of lieutenant colonel. At that time, he became the second-in-command of a new unit comprised of American Loyalists and British cavalry. Known as the British Legion, the force took part in the defense of Savannah in 1779 and the capture of Charleston in 1780. By the time of the latter engagement, Tarleton had assumed command of the British Legion.

With Charleston secured, Tarleton received orders to eradicate any pockets of American resistance, including that offered by Francis Marion. Marion proved too elusive, however, and Tarleton never succeeded in bringing him to bay. To make things worse for Tarleton, he went on to lose the Battle of Cowpens in January 1781, and would eventually become a prisoner of war as part of the British surrender at Yorktown in October of that year.

8. What were some of Marion's major successes?

From beginning to end, Marion's career as an independent commander met with great success. For example, his initial attack resulted in the rout of a detachment of Loyalists without a single American casualty. A few weeks later, Marion hit a British force guarding American prisoners captured at the Battle of Camden, freeing 150 of them. Soon, however, Marion began to fight on a larger scale.

As we have previously seen, after the debacle at Camden the Continental Congress had authorized Nathaniel Greene to take command in the southern states, and George Washington had given him a detachment of his regular soldiers. Upon arriving in the South, Greene had called upon Marion to join him in a campaign to drive the British out of Charleston. In a series of engagements both large and small, Greene, Marion, and other American officers forced the British to withdraw their forces to a perimeter barely encompassing that city. There things stood until the British recognized American independence in 1783.

9. Apparently, there is a picture of Francis Marion in the U.S. Capitol. Can you tell us about the picture and what it entails?

During his guerilla operations, Marion had captured British soldiers, but had also lost some of his own men. At that time in history, opposing forces usually exchanged prisoners of war rather than incarcerating them. To arrange such a trade, in 1781 a British officer journeyed under escort to Marion's hide out on Snow Island. Once Marion and the British officer agreed on terms, legend has it that Marion then offered his counterpart a breakfast consisting of baked sweet potatoes.

Supposedly, the British officer was so impressed with the dedication of Marion and his men in the face of adversity that he immediately switched sides and offered his services to the American cause. In 1820, the American artist John Blake White painted a depiction of this scene, and his work now is on display in the U.S. Capitol.

10. Was he commissioned a general at the end of his career? Or what was his final rank?

In the summer of 1781, the British created a plan by which they would unite separated forces into a unit capable of controlling a key crossroad in the South Carolina hinterland. Marion received orders to prevent that from happening, and to accomplish that task he devised an ingenious strategy to ambush the British forces before they could unite. With a minimum number of casualties, Marion inflicted a decisive defeat on the British. For his actions, Marion received a promotion to the rank of brigadier general. He retained that rank until he tendered his resignation from the army.

11. What have I neglected to ask about this great American hero?

After the war, Marion became a member of the South Carolina Senate. While serving in that body, he argued against punitive treatment for the South Carolinians who had remained loyal to the British during the Revolutionary War. Largely through his efforts, no reprisals took place in that state. Through his wise counsel, the wounds of war thus healed more quickly in South Carolina than in many other states. In this manner, Marion proved to be a great leader in peacetime as well as wartime.

RICHARD MONTGOMERY

Print collection, New York Public Library
Image, public domain

SIXTEEN

Richard Montgomery

MARTYR OF THE AMERICAN REVOLUTION ★ (1738–1775)

---

1. One of the most outstanding generals to serve with George Washington was Richard Montgomery. We know he was born in Ireland; but what else do we know about his early childhood and how did he make it to America?

Richard Montgomery was born in Swords, County Dublin, Ireland on December 2, 1738. Unlike most residents of Ireland at that time, the Montgomery's family did not practice Roman Catholicism. This stems from the fact that the Montgomery family had its roots in Scotland. After England's King Henry VIII created a Church of England independent of papal control in 1536, many residents of Scotland also decided to leave the Roman Catholic Church. However, to a large extent, the Irish chose to remain loyal to the Catholic faith.

These two belief systems came into conflict in Ireland in the early 1600s, when England's King James I began to resettle Scots in Ireland, primarily the northern part of that country. On a side note, many of these settlers would later move to America, and became known as the Scotch-Irish (or Scots-Irish). Some of the Scottish immigrants, including the Montgomery family, became the landed gentry in Ireland. As a consequence, Richard Montgomery's father had the resources to send him to a private school in Leixlip.

In 1754, he became a student at Trinity College in Dublin, Ireland. Before he earned a degree, his family decided that he should join the British Army (as had a number of his ancestors). At the time, wealthy individuals could purchase military commissions from the British government (a practice that continued until 1871). Montgomery's father followed that custom, and bought his son a commission as an ensign in an infantry regiment in 1756.

2. Apparently, he served in the French and Indian War which was also known at the Seven Year's War (1754–1763). Could you briefly provide an overview of that specific war: causes, time frame, and Montgomery's role in it?

By the time Richard Montgomery joined the British Army, a war had broken out between Great Britain and France. In many respects, this war represented merely a continuation of a conflict that had begun in 1689. In that year, King William had declared war on France, primarily to prevent French domination of the continent of Europe. Known in Europe as The War of the League of Augsburg, the conflict had lasted until 1697. Five years later, war again broke out. This conflict, known as The War of the Spanish Succession, lasted until 1713. Thirty years later, the two sides would fight each other again during the War of the Austrian Succession (1740–1748).

When that war ended in 1748, few thought that the two nations would remain at peace for long. These suspicions proved correct, as less than a decade later another war occurred. This conflict differed from its predecessors in one important respect in that it began with an incident that happened in North America. A British attempt to gain control of the Ohio Territory had resulted in a crushing defeat at the hands of the French and their Indian allies (see also the chapter on Horatio Gates).

This defeat had prompted the British to go to war with France. Known in Europe as the Seven Years War, this conflict is referred to as the French and Indian War in the history of the United States. This war brought Montgomery and his regiment to North America in 1757. Montgomery fought bravely in an assault on Fort Carillon (known today as Fort Ticonderoga and situated on shores of the southern tip of Lake Champlain on the border of New York and Vermont), and he earned a promotion. His regiment then participated in a campaign to capture Montreal; one conducted so efficiently that the French offered no resistance. Interestingly, Montgomery would later conduct a similar campaign against the British during the American Revolution.

After participating in a campaign in the Caribbean (where he again served with distinction), Montgomery and his regiment did garrison duty in New York. While there, Montgomery met Janet Livingston, the daughter of a prominent landowner. Initially, this meeting did not result in a romance, as Montgomery and his regiment returned to Great Britain in 1764 after the French and Indian War. But Montgomery increasingly saw the British treatment of the colonies as inherently unjust, and eventually chose to resign his commission in 1772 and return to New York.

Once he arrived there, he purchased a farm north of New York City, resumed his acquaintance with Janet Livingston, and eventually wed her in 1773. He settled into the life of a country gentleman and tried to avoid politics. However, but New York called for a provincial assembly in May of 1775 after the April 19th Battles of Lexington and Concord, Montgomery found himself chosen by his neighbors to serve as a representative. This legislative body soon learned that the Second Continental Congress, which had recently come into session, had taken control of the resistance to Great Britain.

It called on New York to name two generals to serve in the Continental Army, and the assembly nominated Phillip Schuyler and Montgomery. Congress confirmed him as a brigadier general on June 22, 1775.

3. Apparently in one endeavor, Montgomery captured the city of Montreal, as legend has it, without firing a shot! How was this accomplished?

For strategic and diplomatic reasons, the Second Continental Congress had decided at the beginning of the American Revolution to order the Continental Army to capture Canada. Initially appointed as that operation's second in command, Montgomery became the commander as the army moved northward. Crossing into Canada, Montgomery's force captured Fort Chambly and Fort St. John's and then moved on Montreal. Because high ground bordered Montreal, Montgomery proved able to position his artillery in a manner that threatened every part of that city. He then demanded the city's surrender.

Recognizing the futility of trying to resist Montgomery, the British commander acquiesced. Montgomery thus, did force the surrender of Montreal without firing a single shot.

4. Apparently Montgomery actually fought with Benedict Arnold. What are some of the details surrounding that event?

At the same time that Montgomery had advanced with his force into Canada, General Benedict Arnold had begun an operation with the same goal as his objective. Believing that the capture of Quebec would secure control of Canada for the Americans, Arnold and his men had marched directly there. Montgomery, thinking along the same lines as Arnold, moved his troops down the St. Lawrence River towards Quebec after he captured Montreal. Once the two forces effected a rendezvous in early December of 1775, Arnold placed his troops under the command of Montgomery. Montgomery then began a campaign to capture Quebec and secure Canada for the American cause.

5. The 1786 painting by John Turnbull portraying the death of Richard Montgomery is surely a classic (*The Death of General Montgomery in the Attack on Quebec, December 31, 1775*). What do historians actually know about the death of Richard Montgomery and his valor?

Initially, Montgomery hoped that he could simply convince the British to surrender Quebec because he had a numerically superior force. The British commander, Sir Guy Carleton, refused his demand, however, informing Montgomery that he would have to take the city by force. Accordingly, Montgomery then used the artillery he had brought with him to try to bombard the city into surrendering. Unfortunately for Montgomery, he did not possess the fire power necessary to force the British to capitulate. Montgomery decided that he had no option left except an all-out assault of the city. He waited until a snowstorm on the night of December 30, and then ordered his troops to attack the city.

Initially, the Americans met with success, advancing towards the center of the city with Montgomery leading the assault. But just as success seemed within the grasp of the Americans, grapeshot fired by a British cannon fatally wounded Montgomery. With his death, the Montgomery's assault came to an end, and the Americans soon began a retreat. Continental Army troops fled so precipitously that they abandoned Montgomery's body. The next morning, an American soldier captured by the British identified his corpse.

6. Like Ulysses S. Grant, Civil War general and president, who is buried in Grant's Tomb in New York City, Montgomery's remains apparently are also in New York City. Where exactly was he buried?

After his death, the British commander had Montgomery buried with military honors in Quebec. There his remains lay until 1818. At that time, the governor of New York asked the British for permission to exhume Montgomery's body and move it to American soil. Earlier that year, the British and American governments had negotiated a treaty that had solved a number of problems that had driven the two countries to war in 1812, and as a result of the easing of tensions the British agreed to the governor's request. As a consequence, Montgomery's body was moved in the summer of 1818 to St. Paul's Chapel in New York City. A monument honoring him had been built there in 1776, and the church members buried his remains near that memorial; where they are today.

DANIEL MORGAN

By Charles Wilson Peale, c. 1794
Independence Hall, Philadelphia (INDE 11844)
Image, public domain

SEVENTEEN

Daniel Morgan

"OLD WAGONER" ★ (1736–1802)

---

1. Daniel Morgan, one of Washington's generals, was actually a first cousin of Daniel Boone. Where and when was he born?

Although most people associate Daniel Morgan with either Virginia or West Virginia, he actually was born in New Jersey on July 6, 1736. Accounts vary, but most historians feel that he was born in Lebanon Township; however, they have found no definitive proof to support that assertion. In a similar vein, historians have mixed opinions regarding the widely-held belief that Daniel Boone and Daniel Morgan were cousins.

This assumption apparently came from the fact that Daniel Boone's mother had the last name of Morgan. Recently, a genealogist supplied evidence that seems to prove that no such family tie existed, but like many urban myths the supposed relationship between Morgan and Boone will undoubtedly live on.

2. How did Morgan first get involved in the military?

From all accounts, at the age of 17, Morgan had an unpleasant disagreement with his father, and decided to strike out on his own. After a brief sojourn in Pennsylvania, Morgan purchased a farm in Virginia. Soon thereafter, he became a teamster for a British military expedition into the Ohio Territory under the command of General Edward Braddock during the French and Indian War. After Braddock's defeat south of present-day Pittsburgh, the remnants of his army, including Morgan, retreated. In route, Morgan became angry with a British officer, who then struck him with the flat of his sword. Morgan then assaulted him, and the officer retaliated by sentencing him to 500 lashes. He later boasted that the British miscounted and only gave him 499!

While many individuals died from such a punishment, Morgan somehow survived, but the incident would give him a life-long hatred of the British. Braddock's campaign led to the start of a wide-spread conflict that became known in British North America as the French and Indian War and elsewhere as the Seven Years' War, and after recovering from his lashing Morgan decided to volunteer for service in a Virginia regiment. Although still a young man, Morgan received an appointment as a subordinate officer.

Unfortunately, Morgan's story at that point once again becomes problematic. According to popular legend, Morgan led a spirited defense of Fort Edward in western Virginia in 1757. However, no record mentions a battle that took place at Fort Edward in that year. What historians have learned is that in 1757 Morgan received dispatches to deliver to a frontier outpost, and in route Native Americans ambushed his party. During the encounter, a bullet hit him from behind. It entered his neck, and took out a row of teeth as it exited through his mouth. Here again, Morgan miraculously recovered, and lived to fight another day.

3. Apparently he was associated with marksmanship, and his men distinguished themselves in this regard on several occasions. What do we know about this?

At the start of the American Revolution, Morgan's county decided to support the Patriot cause, and asked him to become the captain of the men who volunteered for service from that area. In June 1775, nearly 100 men, known as "Morgan's Riflemen," followed him to the Continental Army's encampment outside Boston. His men brought the weapons from home, and most of them carried rifles. This type of weapon had a groove, known as rifling, engraved in its barrel. That feature gave the weapon far greater accuracy than the musket most soldiers carried, and this made his unit highly sought after as sharpshooters. His men also wore distinctive dress in that they wore hunting shirts—something specifically American. The sight of these men with their known accuracy caused great fear in the British.

Indeed, he would employ his men as what we would call snipers during the rest of the campaign to drive the British out of Boston. In 1777, George Washington would give him command of the Provisional Rifle Corps, a unit made up of 500 men from Virginia, Pennsylvania, and Maryland, who all carried rifles. They would prove extremely useful in battle that fall.

4. Morgan fought alongside Benedict Arnold on a number of occasions. What do we know about his endeavors?

At the beginning of the American Revolution, the Continental Congress decided that for strategic reasons it should order George Washington to allocate troops under his command for a campaign to capture Canada. Accordingly, in the summer of 1775 Washington tasked General Richard Montgomery with this assignment. After Montgomery's departure, a colonel from Connecticut named Benedict Arnold suggested to Washington that he should lead another force into Canada to double the chances of capturing that colony. Washington found Arnold's logic compelling, and authorized him to take 1,000 soldiers northward. Marching through present-day Maine, Arnold's force reached Canada in November, and alongside Montgomery attacked Quebec on New Year's Eve 1775.

Daniel Morgan commanded a company in Arnold's expedition, and personally led one wing of Arnold's assault force at the Battle of Quebec. Showing remarkable bravery, Morgan personally led his men forward. Remarkably, he suffered a leg wound and the British succeeded in capturing him. Exchanged through a prisoner-of-war cartel arranged between the British and Americans, Morgan's release was followed by promotion to colonel of the 11th Virginia Regiment. He rejoined the army in 1777, and throughout the year he led troops using hit-and-run tactics against the British in New Jersey and New York. Later in 1777 he was assigned to General Horatio Gates and participated in the key Battle of Saratoga.

5. The literature seems to suggest he was not a politician, and was passed over many times for the rank of general. Can you tell us a bit about his difficult rise to general?

At the start of the American Revolution, Morgan had found the rank of captain perfectly acceptable. In fact, it appears that his promotion to colonel for his bravery in the assault on Quebec took him completely by surprise. His estimation of his talents grew over the next few years, however, as on a number of occasions his superiors gave him important command responsibilities.

Despite his successes, Morgan remained at the rank of colonel into the summer of 1779. Having no one in Congress to champion the case for promoting him, Morgan tendered his resignation from the Continental Army on June 30, 1779. He would remain a civilian for over a year before rejoining the Patriot cause.

6. What do we know about his earlier role in the Battle of Saratoga?

In June of 1777, the British general John Burgoyne led a large force out of Canada as part of a plan to split New England off from the rest of the United States. This threat did not fully sink in for the

Americans until Burgoyne recaptured Fort Ticonderoga. That loss caused Congress to begin gradually shifting forces northward to deal with the advancing British. In August 1777, Morgan received orders to take his company to New York to become a part of the Northern Department of the Continental Army under the command of General Horatio Gates.

By the middle of September, Gates had occupied a strong defensive position a few miles outside of Saratoga, New York, and waited for Burgoyne to arrive. When Burgoyne's force finally advanced, Benedict Arnold (by now Gates's second-in-command) suggested moving Morgan's Provisional Rifle Corps to a wooded area bordering a farm owned by a man named Freeman. Gates agreed, and when Morgan advanced into the copse of trees he saw a British force moving across the field. Immediately he ordered his men to fire on the British, instructing them to aim at the British officers.

Within minutes, every British officer became a casualty, throwing the rank and file soldiers into confusion. Morgan then ordered his men to charge, and they quickly drove the British from the field. Unfortunately for Morgan, he then ran into the main body of that wing of the British army, who drove the Americans back to the wooded area. A general engagement then developed that would last the rest of the day. By the end of the battle, the British had driven the Americans back to their original position, but had suffered twice as many casualties.

A few weeks later, Burgoyne tried again to work his way southward towards Albany. Anticipating this strategy, Gates deployed his army to block Burgoyne, giving Morgan command of the left wing. As fate would have it, that put Morgan squarely in the path that Burgoyne chose for his assault. But Morgan skillfully positioned his riflemen and a New Hampshire regiment in such a manner that the two American forces proved able to catch the advancing British soldiers in a crossfire. Here again, officers became the primary targets, and eventually one

American sharpshooter succeeded in killing the British officer leading the attack, Brigadier General Simon Fraser. With their commander dead, the British soon retreated, and Morgan and his men drove them back to their original starting point.

This engagement, known as the Battle of Bemis Heights, sealed the fate of Burgoyne's campaign, convincing him that he had no other recourse than surrender. As a tribute to his contributions to the American victory, the official painting of the surrender at Saratoga puts Morgan in close proximity to Gates. By contrast, Benedict Arnold, who also provided a significant contribution to the victory at Saratoga, does not appear in the painting.

7. After General Gates's disaster at the Battle of Camden on August 16, 1780, Morgan was again thrust into service, or decided to return to battle. Which was it? And what was his role at that time?

The year 1780 started inauspiciously for the Patriot cause when a British campaign succeeded in capturing Charleston, South Carolina, the largest city in the South. The situation worsened when an American force under the command of Horatio Gates, sent south by the Continental Congress to rectify the situation, suffered a disastrous defeat at the Battle of Camden. At that point in time, Morgan decided to re-enter the fray, and journeyed to Hillsborough, North Carolina, to offer his services to Gates. Soon thereafter, Congress replaced Gates with Nathaniel Greene.

Giving Morgan 600 men, Greene ordered him to take his force into northwestern South Carolina. Along the way, Morgan picked up a few more volunteers. Made aware of this foray, Lord Cornwallis, the British Commander in that theater of operations, deployed a force to move quickly against Morgan. When he learned of the advancing British force, Morgan chose to stand and fight. At first glance, he seemed to have made a serious miscalculation when selecting a spot to make his stand. That location, a grazing area known as Cowpens, had a river running behind it. This meant that if they suffered a defeat, the Americans would find retreat nearly impossible.

He then seemingly made a second mistake when he put untested militiamen in his front rank. However, it soon turned out that Morgan had deliberately arranged his forces that way. Unbeknownst to the British, Morgan had placed a few sharpshooters in front of the militia. He asked his sharpshooters and militia to fire two rounds at the advancing British, and told them that they could then retreat. At Camden, the militia present that day had ignominiously fled the battlefield, and Morgan sensed that his strategy for the engagement at Cowpens would convince the British that the same thing was happening again.

Events on January 17, 1781 at Cowpens proved Morgan correct. Believing that they had routed Morgan when they saw the retreating militiamen, the British soldiers broke ranks to pursue them. But as the British reached the crest of a small hill, Morgan had his Patriot soldiers rise up from hidden positions to deliver a punishing volley at point-blank range. The American force then retreated, and the surviving British soldiers pursued them. Suddenly, the Americans turned and fired a second volley. Taken completely by surprise, the British soldiers wavered.

At that moment, Morgan ordered his 100 cavalrymen to charge the British in a pincer movement. Thoroughly routed, the surviving British soldiers attempted to flee, but Morgan's force captured over 800 of them. By the end of the day, Morgan had inflicted one of the worst defeats on the British that they would suffer during the entire course of the war. Soon after his victory, Morgan finally received a promotion to the rank of brigadier general.

8. Apparently, before he died he also served our country in a few other capacities. What were some of them?

After he resigned for a second and final time from the Continental Army a few weeks after the Battle of Cowpens, Morgan returned to his farm in Virginia. In 1794, however, Morgan would return to military service when George Washington (now the president) felt the need for a show of force against farmers in western Pennsylvania who had refused to pay taxes to the federal government. Receiving a promotion to major general, Morgan took command of one wing of the army as it advanced against the farmers.

After the resistance, known in American history as the Whiskey Rebellion, dissolved, Washington ordered a portion of the army to remain in Pennsylvania, and gave Morgan command of these troops. Once he received his discharge, Morgan ran for a seat in the U.S. House of Representatives. Easily winning the election, Morgan served one term in Congress. He did not seek reelection in 1798, and died four years later. Although largely forgotten today, Daniel Morgan's name lives on in numerous states that named towns and counties after him.

WILLIAM MOULTRIE

By Charles Wilson Peale, 1782
National Portrait Gallery, Washington, D.C.
Image, public domain

EIGHTEEN

William Moultrie

SOUTHERN PATRIOT ★ (1730–1805)

1. Many individuals such as William Moultrie served George Washington during the Revolutionary War. Where was he born and what do we know about his background?

William Moultrie was born on November 23, 1730, in Charleston, South Carolina. His father practiced medicine and had a small plantation. These endeavors allowed him to give his son an affluent upbringing. Although no records prove it, it would appear from his personal papers and correspondence that Moultrie must have received an excellent education while growing up, as he demonstrated a literary fluency during his adult years.

At the age of 19, Moultrie married Damaris Elizabeth de St. Julien, a wealthy descendant of French immigrants, a young lady whose dowry included an 8,000-acre plantation. Through the oversight of his wife's resources (required by English law at the time), he also purchased a 1000-acre plantation from his brother-in-law.

Soon after his wedding, Moultrie's neighbors elected him to the South Carolina House of Commons. In 1758, while the northern British-American colonies participated in the French and Indian War (also known as the Seven Years' War), South Carolina became embroiled in a conflict against the Cherokee tribe living in the western part of the colony. Moultrie served as a captain in the South Carolina militia, going on active duty against the Cherokees in 1760.

After the 1758–1761 Anglo-Cherokee War ended, Moultrie remained active in the South Carolina militia, receiving a promotion to the rank of colonel in 1774. One year later, he would become a colonel in a much different military organization—one bent on resisting the English rather than fighting alongside them.

2. He prevented Sir Henry Clinton and Sir Peter Parker from taking a fort, in effect saving Charleston from the British. The fort was later named after him. What do we know about his endeavors in that action?

With the growing tension between the colonies and Great Britain over taxation without representation, South Carolina decided to support opposition to the British policies. Accordingly, South Carolina chose Moultrie to represent it in the Continental Congress. Moultrie declined the honor, choosing instead to remain in South Carolina. His decision soon paid dividends, as Moultrie received a colonel's commission and command of the 2nd South Carolina Regiment after news of the Battles at Lexington and Concord (April 19, 1775) in Massachusetts reached the colony.

In this capacity, in March of 1776 he oversaw the construction of a fort on Sullivan's Island, which guarded the seaward approaches to Charleston. His preparedness proved prescient in the spring of that year, when the British began a campaign to capture the city. To implement this plan, a fleet of nine ships under the command of Admiral Peter Parker, mounting 270 cannon and carrying an invasion force of 2,200 soldiers, sailed into Charleston Harbor in early June 1776. As a first step, the fleet landed the soldiers, under the command of General Henry Clinton, on Long Island, just north of Sullivan's Island. In the ensuing operation, the soldiers advanced on Moultrie's fort, while at the same time the British ships attacked the installation from the seaward side.

Moultrie could only oppose the combined British force with 31 cannons of his own, and for that reason few gave his 450 defenders a chance of repulsing the British. Amazingly, they did just that. A primary reason for the American victory that day stemmed from the fact that the British naval bombardment proved ineffective, as cannon balls that would have pulverized a masonry fort simply buried themselves in the sand and palmetto logs that comprised Moultrie's structure. In contrast, Moultrie's artillerymen provided accurate fire throughout the battle. Parker's flagship, the HMS *Bristol*, for example, was hit 70 times, and suffered 111 casualties.

The British failure on June 28, 1776 also resulted from the inability of Clinton to move his force across the channel that separated Long Island from Sullivan's Island. Anticipating such a move, Moultrie had strategically located some of his artillery pieces to command that passage, and after a brief foray Clinton had given up the effort. Because of these factors, the Americans handed the British a humiliating defeat, made worse for them a few weeks later when they learned that the Continental Congress had issued the Declaration of Independence. Moultrie's efforts had prevented the British from landing forces in Charleston by land and sea.

Deciding that they did not have a force of sufficient force to take Charleston, the British withdrew in July. In recognition of his stalwart defense of Charleston, days later the South Carolina legislature decided to honor Moultrie by naming the fort for him.

3. Legend has it that Moultrie had his own flag with the word "Liberty" emblazoned on it. What is the story of this legend and this flag, and does it still fly anywhere?

When the Revolutionary War started, every unit that fought against the British did so under banners of their own creation. In South Carolina, provincial leaders gave Moultrie the task of creating a flag for the colony's soldiers. In choosing a design, Moultrie remembered that South Carolinians had showed their opposition to the Stamp Act in 1765 by protesting under a blue banner that featured three crescents in the upper left hand corner. To this basic design, Moultrie made two adjustments. First, he reduced the number of crescents from three down to one. And second, he had the word "Liberty" embroidered into the crescent. This flag flew over his fort on Sullivan's Island in June of 1776. A British shell struck the flag post during the battle, causing it to fall to the ground.

Bravely, Sergeant William Jasper grabbed the flag and nailed it to the remnant of the flagpole. Firmly secured, the flag then flew over the fort for the rest of the battle. Years later, when South Carolina designed a state flag, it adopted the key elements of Moultrie's creation. South Carolina's flag differed only from the addition of a palmetto tree and the subtraction of the word "Liberty" from the crescent.

4. Sadly, Moultire's skills and courage could not prevent the British from eventually taking Savannah, Georgia, and Charleston. What transpired in those campaigns?

After their repulse at Charleston in 1776, the British had once again moved south in 1778. Instead of returning to Charleston, however, the British decided to move on Savannah, Georgia. At first, it appeared that an American force might be able to defend the city, but information given to the British commander by a slave allowed him to move a significant number of his men around the American flank. In the ensuing battle, the British took advantage of their superior position, and routed the defenders. That allowed them to take possession of the city at a comparatively small cost.

Made aware of the British campaign, the Continental Congress had placed General Benjamin Lincoln (1773–1810) in command of the Department of the South and had ordered him to defend Savannah; by the time he arrived in the vicinity, however, the city had already fallen. But when the British attempted to follow up their victory by taking possession of nearby Port Royal Island, South Carolina, a month later, the newly-arrived American general responded by ordering troops under his control to defend the island. Lincoln gave command of this tactical force to William Moultrie (promoted to brigadier general after the Battle of Sullivan's Island).

On February 2, 1779, he engaged a British detachment roughly equal to his size on the island, and by the end of the day held the field of battle. This victory gave the British pause, and allowed a combined Franco-American force to try to recapture Savannah later that year. That effort failed, however, and a year later the British began a campaign designed to capture Charleston. This time, in May 1780, the British succeeded, forcing the 5,200 defenders of the city (including Lincoln and Moultrie) to surrender. Moultrie remained a British captive until a prisoner-of-war exchange returned him to the Continental Army in 1782. When he rejoined the army, Congress promoted him to major general—the last person to receive that rank during the American Revolution.

Interestingly, two of Moultrie's brothers were Loyalists and knowing this, the British at one point offered Moultrie freedom, a British colonelcy, and command of a British regiment in Jamaica if he would become a traitor and leave the Patriot cause. This was an offer he adamantly refused.

5. What is the legacy of William Moultrie?

Like many of his compatriots, Moultrie gave of himself after the conflict ended, serving two terms as the governor of South Carolina. As previously noted, he also contributed to his state by providing the banner that South Carolina used as a template for creating its flag. Moultrie also wrote and published in 1802 *Memoirs of the Revolution as far as it Related to the States of North and South Carolina.* His work was one of the most widely-read and cited first-hand accounts of the war for American independence.

But his most lasting legacy was the fort that he built on Sullivan's Island. After the American Revolution, the U.S. Army kept it as a military installation, and in December of 1860 a small detachment of soldiers under the command of Major Robert Anderson occupied Fort Moultrie. That led to a critical moment in American history, because South Carolina seceded from the Union at that point in time, and upon proclaiming its independence the state demanded that Anderson surrender the fort.

Rather than capitulate, Anderson chose instead to abandon Fort Moultrie and transfer his force to the more defensible Fort Sumter, located on an island in the middle of Charleston harbor. Because of Anderson's action, the U.S. flag thus still flew over Charleston harbor when Abraham Lincoln became president in March of 1861. Lincoln chose to keep possession of Fort Sumter, and the newly created Confederate States of America decided to fire on the installation—an act that started the Civil War.

Had Anderson not evacuated Fort Moultrie and moved his force to Fort Sumter, the entire course of American history would have undoubtedly been different. Creating a fort that became a potential flash point, then, stands as William Moultrie's most lasting legacy in American history.

ENOCH POOR

Unknown engraver
Image, public domain

NINETEEN

Enoch Poor

FROM PRIVATE TO GENERAL ★ (1736–1780)

1. One of "Washington's Generals" was Enoch Poor, who was described by Washington as "an officer of distinguished merit, one who as a citizen and soldier had every claim to the esteem and regard of his country." Where was he born, and what do we know about his early life?

Enoch Poor was born on June 21, 1736 in Andover, Massachusetts. His father, Thomas Poor, volunteered for military service in 1745, during a conflict known in American History as King George's War. Not long after Thomas enlisted, he became part of a colonial expedition that captured the French fortress at Louisburg in Nova Scotia. In the peace treaty that ended King George's War, the British government agreed to return control of Louisburg back to the French—an action that exceedingly vexed the colonies that had captured it in 1745.

Ten years later, during the French and Indian War a force of British regulars and colonial militia, including Enoch Poor who had enlisted as a private in a Massachusetts unit, captured Louisburg for a second time. After five years of service in the Massachusetts militia, Poor left the military and returned home to Andover. There, he fell in love with a young lady named Martha Osgood. By all accounts, she felt the same way about him. Unfortunately, Martha's father did not deem

Poor a suitable match for his daughter, and refused to give his consent to a marriage of the two.

Indeed, he went so far as to order his daughter to stay in her second-floor bedroom. Undeterred, Poor procured a ladder, and whisked Martha away. The newlyweds then settled in Exeter, New Hampshire to begin their lives as husband and wife.

2. Apparently one of his early skills was as a shipbuilder. How did this help him in the American Revolution?

As many parents did in those days, Thomas Poor secured an apprenticeship for Enoch, choosing the carpenter's trade for his son. As evidenced by a piece of furniture that he made, now in the possession of a private collector, Poor learned his craft well. At some point in time, Poor translated his woodworking skills into a talent for shipbuilding. By the time of the American Revolution, Poor owned a thriving shipbuilding business in Exeter, and he used that skill to offer a unique type of potential help for the Patriot cause after the Battles of Lexington and Concord. Fearing that the British might attack the port city of Portsmouth, New Hampshire, Poor's firm built vessels known as "fire ships."

If the British Navy ever did menace the city, the defenders of Portsmouth would light the vessels on fire, and would sail them in the direction of the British fleet. The threat of immolation from these vessels would cause the British fleet to break off its attack and head for safety out in the open seas. Due in part to this threat, the British never attacked Portsmouth.

3. When the "Stamp Act" was enacted by the Parliament of Great Britain in 1765, Poor apparently was at the forefront of opposition. What was his involvement?

Many American colonists resented British attempts after the French and Indian War to tax the colonies. They opposed these efforts largely for two reasons. First, no colonists sat in Parliament, which meant that the Americans had no say in the levying of taxes. And second, many colonists asserted that they had paid taxes within their own colonies to support the British war effort, making the British taxes excessive. Poor had a third reason for opposing the British taxes; he had risked his life for the British during the French and Indian War.

For those reasons, Poor became quite vocal in his opposition to the British policies, and became a member of a number of local committees created to adopt strategies for resisting the British efforts to tax the colonies. Poor transitioned to opposing the British militarily in the spring of 1775, when the New Hampshire provisional government gave him command of its Second Infantry Regiment after the outbreak of hostilities. Initially deployed to defend Portsmouth, in June 1775, Poor's regiment received orders to join Washington's army at its encampment outside Boston.

4. Poor later spent the winter of 1777–1778 at Valley Forge, Pennsylvania. What do we know if about his activities prior to that time?

After joining Washington's army. Poor received orders to join a campaign under the command of General Phillip Schuyler that Congress had ordered to take possession of Canada. Command of the invasion force soon passed to General Richard Montgomery. This force captured British outposts, including Montreal, in the late fall and early winter of 1775. In November, Montgomery arrived outside Quebec, and found another American force under the command of Benedict Arnold already deployed there. The two generals joined forces, and attacked Quebec on New Year's Eve, 1775.

Initially successful, the American attack lost momentum after the British defenders killed Montgomery and wounded Arnold. Retreating from Quebec, the Americans suffered another defeat at the Battle of Trois-Rivieres in May 1776. Eventually, the remnants of the American invasion force made their way back to Washington's army, rejoining it after it had found temporary sanctuary in Pennsylvania in December 1776.

Assigned to a brigade commanded by General Arthur St. Clair, Poor's men then took part in the Battle of Trenton on December 26, 1776. A few days later, on January 3, 1777, Poor and his men helped Washington win the Battle of Princeton. For the outstanding service that he had provided to that point in time, Poor received a promotion to the rank of brigadier general in February 1777. In the summer of that year, Washington sent Poor to join the northern army commanded by General Horatio Gates.

In perhaps one of his finest moments, Poor provided exceptional battlefield leadership at the Battles of Freeman's Farm (September 19, 1777) and Bemis Heights (October 7, 1777), two engagements that allowed Gates to force the surrender of a British-Hessian force commanded by General John Burgoyne at Saratoga. After the capitulation, Poor and his men rejoined Washington's army at its winter camp at Valley Forge for the harsh winter of 1777–1778..

5. He was involved in the June 1778 Battle of Monmouth. What leadership did he provide there?

After capturing Philadelphia in the fall of 1777, the British had soon realized that they had gained no strategic advantage from that accomplishment. For that reason, the British chose to abandon the city the following spring, and move their base of operations to New York City. Learning of the British withdrawal, Washington planned an attack on them while their force lay stretched out on the route to New York

City. This attack took place near Monmouth Court House in New Jersey in June of 1778.

Almost from the first shot fired that day, Washington's battle plan went awry, and it looked like the Americans would suffer an embarrassing defeat. But Washington rode to the front, took personal charge of the American forces, and reversed the tide of battle. In fact, he felt so confident of victory that late in the day he ordered Poor, by now a brigade commander, to use his force to assault the British right flank.

Darkness prevented Poor from engaging the British, and during the night the British left the battlefield and resumed their march to New York. Although some historians consider the Battle of Monmouth a draw, most think that Washington should receive credit for a victory for holding the field of battle.

6. There seems to be some disagreement about Poor's death. What do historians know in this regard?

Poor had served under the command of General John Sullivan in the abortive Canadian campaign in 1775–1776, and in 1779–1780 he once again joined an expedition led by the New Hampshire native. During the latter campaign, Poor helped Sullivan win a decisive victory over a force comprised of Iroquois warriors, Loyalists, and approximately 60 British regulars at the Battle of Newtown. After the campaign ended, Poor received orders to report for duty in New Jersey. There, he died on September 8, 1780. However, accounts differ regarding the cause of his death.

At the time of his death, the physician who attended to him listed typhus as the cause. One hundred years later, however, an alternative theory of his death emerged. According to a paper delivered to the Massachusetts Historical Society, Poor angered a subordinate officer from Massachusetts named John Porter, who challenged the general to a duel. When the two faced each other, Porter inflicted a wound that proved fatal. Most historians discount this story, pointing to the fact that there are first-hand accounts of Poor's death from natural causes. The naysayers also point to the fact that Porter before the war had been a minister, making it unlikely that he would engage in a duel. Despite all the evidence to the contrary, however, rumors still persist regarding Poor's death.

His funeral was attended by both Washington and Lafayette; perhaps no greater honor could be afforded a soldier. A number of generals in the American Army died during the American Revolution. Some, like the Baron de Kalb, died in battle, while others, such as John Thomas, succumbed to natural causes. While Washington mourned them all, he (along with Lafayette) seemed particularly saddened by the death of Poor. Indeed, the eulogies that the two gave him contained words of the highest praise. Clearly, Poor impressed his peers and superiors in a way that most histories of the Revolutionary War era do not properly convey.

7. Where is he buried and, isn't there a statue honoring him in New Jersey?

During the American Civil War, it became common practice to embalm the corpses of high-ranking officers, and then return their remains to their home-towns. At the time of the American Revolution, however, the capability to preserve remains did not exist. Therefore, most generals that perished during the conflict were buried where they died. So rather than bury him in Exeter, New Hampshire, Poor's comrades chose to make Hackensack, New Jersey, his final resting place.

Only a few feet from his grave stands a monument in his honor. The Sons of the American Revolution placed a plaque on it, recognizing that Poor "secured the respect of all who were under his command, gained for all time the esteem of his fellow officers, and the confidence of Washington and Lafayette." This monument in Bergen County, New Jersey thus perfectly captures the essence of a general who ably served the cause, but unfortunately did not live long enough to see the nation win its independence.

CASIMIR PULASKI

By James Hopwood, engraver
National Park Service
Image, public domain

TWENTY

Casimir Pulaski

FATHER OF THE AMERICAN CAVALRY ★ (1745–1779)

1. Casimir Pulaski served bravely in the American Revolution. Where was he born? Didn't he have a massive amount of experience before coming to assist George Washington and the American Revolution?. Can you give us an overview summary?

Casimir Pulaski was born into a noble family in Warsaw, Poland on March 6, 1745. His father held a number of important leadership positions in and around Warsaw. Afforded the opportunity to attend a local college operated by a Roman Catholic religious order, Pulaski left school before completing his degree. Instead, he chose to enter the political realm, becoming a page of the Duke of Courland. After six months, Pulaski's father gave him a position of authority in the town of Zezulińce in Padole.

In 1767, feeling that Poland had become little more than a puppet state of Russia, Pulaski became involved in a movement to curtail the influence of the Russians in his country—a Polish anti-Russian insurrection that from 1768–1772 was known as the Confederation of Bar . Given the rank of colonel and put in command of a cavalry regiment, Pulaski fought his first battle in March of that year. While his initial engagements went well, the Russians eventually forced him to surrender in June.

Released on his promise to never again take up arms against his captors, Pulaski soon went back on his word and resumed the fight against the Russians. Over the next five years, Pulaski emerged as a brave, but occasionally foolhardy, cavalry commander. Although his side won some victories, eventually the Russians forced the Poles to sue for peace. Rather than submit to Russian dominance and facing potential charges of attempted regicide, Pulaski fled into exile. After brief stays in Prussia and the Ottoman Empire during the Russo-Turkish War, Pulaski eventually made his way to France. There, he tried to enlist in the French Army, but his efforts proved unsuccessful. He was unsuccessful with the French because during the years of rebellion in Poland, Pulaski had participated in an attempt to kidnap the pro-Russian King Stanislaw II Augustus. This failed attempt ended foreign support from Austria and France for Pulaski's cause, the Confederation of Bar and led to its final defeat in 1772. However, while in exile in France in the spring of 1777, he met Benjamin Franklin who was then the American commissioner to France to gain French support for the American Revolution.

Learning of Pulaski's military prowess, Franklin sent a letter to George Washington, suggesting that the American cause could use such an officer. Believing that with Franklin's recommendation he could secure a commission in the Continental Army, Pulaski sailed to America, arriving in Boston, Massachusetts in July 1777.

2. Pulaski once said "I came here, where freedom is being defended to serve it, and to live or die for it." When did he utter these words, and what were the circumstances surrounding it?

After arriving in Massachusetts, Pulaski offered his services to the Patriot cause in a letter to George Washington. In that correspondence, Pulaski wrote the cited quotation. As we have seen, Pulaski had taken up arms in his native Poland to try to rid his homeland of foreign influence, and his words suggest that he now saw the American colonies fighting for the same principles. And, he proved willing to back up his words with his actions.

3. Whether rumor, legend, or perhaps even truth, it is often said that at one point Pulaski saved George Washington's life. What do historians know of that purported incident?

Although extremely well qualified, Pulaski at first did not receive a commission in the Continental Army. After unsuccessfully pleading his case directly to the Continental Congress, Pulaski returned to Washington's army in September 1777 and offered to serve as a volunteer until a commission became available. Shortly after Pulaski arrived in camp, a British force under the commander of General William Howe began an advance on Philadelphia. Moving his army to thwart the British advance, Washington engaged the enemy along the banks of Brandywine Creek in what is commonly known as the Battle of Brandywine (September 11, 1777).

As he had done at the Battle of Long Island (August 27, 1776), Howe feinted an attack on the center of Washington's line while deploying other detachments to make flanking assaults. Howe's plan worked to perfection, and by the end of the day the British had forced the Americans to flee from the battlefield. In the confusion, it appeared that the British might have an opportunity to capture Washington himself. Sensing the danger, Pulaski led a contingent of Washington's mounted guard unit to reconnoiter.

Reporting the dire situation to the general, Pulaski received permission from Washington to use the troops to fight a rear guard action. Although accounts vary, many historians believe that Pulaski's decisive action prevented the British from killing or capturing Washington. As evidence, they point to the fact that soon after the battle, Washington recommended that Congress give Pulaski a commission as a brigadier general in the Continental Army and the Continental Congress agreed.

Immediately proving his worthiness for an appointment at that rank, in the fall of 1777 Pulaski helped keep safe a foraging expedition into southern New Jersey under the command of General Anthony Wayne. For his efforts, Pulaski received a commendation from Wayne. Pulaski thus seemed poised to make a significant contribution to the American cause and much of his leadership as a general was in leading small groups of cavalry in scouting and raiding parties.

4. Pulaski apparently had some difficulty with the English language and also with his high standards that he tried to impose. How did this impact his leadership?

A number of Washington's generals came from other countries, and most of them, including Pulaski, spoke little to no English. This linguistic barrier proved difficult for Pulaski to overcome, and occasionally resulted in communication breakdowns on the battlefield. In addition, from all accounts Pulaski had an irascible temperament, and frequently castigated those around him for mistakes large and small.

Already unhappy with his strained relationship with his subordinates as 1777 drew to a close, Pulaski's mood further soured when he learned that his superiors had rejected his proposal to equip a portion of his cavalrymen with lances Thoroughly frustrated, Pulaski resigned his commission in March 1778, returned to Valley Forge where General Washington was headquartered and once again became a volunteer with Washington's army

5. Pulaski worked with two other generals in the Southern theater of the war. What accomplishments emanated from these partnerships?

After remaining with Washington's army for a short period of time, Pulaski decided to rejoin the fight. To that end, he approached General Horatio Gates with a proposal to raise a cavalry outfit that he would then command. Impressed with Pulaski's plan, Gates passed it on to Congress. That legislative body agreed, and restored Pulaski to the rank of brigadier general. Pulaski then raised a combined force of cavalry and infantry numbering around 250 men. He trained the unit, named the Pulaski Legion, with his Hungarian friend and fellow soldier Michael Kovats (1724–1729), often providing personal funds to support the unit. Apart from his temperamental personality and lack of proficiency in English, another aspect of the longstanding difficulty Pulaski experience also was his requisitioning of horses and supplies from suspected Loyalist sympathizers—a practice common in European warfare but not in line with the revolutionary aims in American colonies.

Pulaski's Legion had limited success in New Jersey in the fall of 1778 in an effort known as the Affair at Little Egg Harbor (October 15, 1778). Soon thereafter, Pulaski received a new set of orders from Washington.

Hoping to neutralize the Native American threat to the frontiers of New York and Pennsylvania, Washington planned a major operation for the spring of 1779 to subdue the Native Americans in those areas. Reluctant to participate in that type of campaign, Pulaski requested that Washington reassign him. Obliging Pulaski, Washington ordered him to instead become part of the Continental Army of the South. Commanded by General Benjamin Lincoln, this force had Charleston, South Carolina as its base of operations. When Pulaski arrived there, he learned of the presence of a British force south of the city.

Suspecting that the British planned an assault on Charleston, Pulaski took his legion out to confront them. In the ensuing battle, Pulaski's force suffered heavy losses, and had to retreat back into the city. In the fall of 1779, French forces under the command of the French admiral Charles Henri Hector, comte d'Estaing (1729–1794) arrived in South Carolina to aid the colonial cause. With 6,000 troops between them, d'Estaing and Lincoln planned an assault on Savannah, Georgia.

Commanding the French and American cavalry, Pulaski reduced a British stronghold near the Ogeechee River, clearing the way for d'Estaing and Lincoln to advance to Savannah. Initially, d'Estaing planned to use siege tactics to force the surrender of the city, but by early October he decided that he could only capture Savannah through an assault. He chose October 9 as the date for the attack, and ordered Pulaski to deploy the cavalry as part of the operation.

6. Almost like the real-life D'Artagnan, the inspiration for the character of the same name in *The Three Musketeers*, Pulaski was killed leading a group of his men. What were the details of his death, and what happened to his body?

During the October 9, 1779 assault on Savannah, the first two attacks on the British defensive line ended in failure. As the French attackers began to retreat, Pulaski tried to turn the tide of battle by launching a cavalry charge. As he rode at the front of his men, an artillery projectile hit him, inflicting a grievous wound. Hoping to save his life, American soldiers took him on board an American vessel *Wasp* for medical care. For years, individuals disagreed about what happened next. Some reports stated that Pulaski died on the ship on between October 11–15 and the ship's crew then buried Pulaski at sea. Other accounts, however, asserted that Pulaski had been taken off the ship and moved to a nearby plantation while still alive. In this telling, Pulaski then died, and the family that owned the plantation buried him there.

Years later, an excavation on the plantation discovered remains that in many ways matched descriptions of Pulaski. In 2019, tests conducted by the Smithsonian Institution proved to most experts' satisfaction that the bones discovered on the plantation were those of Casimir Pulaski. The bones then received a military burial next to a monument in his honor located in Savannah, the city that Casimir Pulaski tried to liberate in 1779.

7. There are all kinds of monuments, highways and other statues honoring this great man. Can you tell us about a few?

Casimir Pulaski lives on today through a number of different memorials to him. As we have seen, Savannah, Georgia, erected a monument to him in the 1850s, while New Jersey named a bridge after him in 1932. In a similar fashion, New York named a bridge over Newtown Creek after him. A city in Tennessee and a county in George both bear his name. The State of Illinois has a holiday in March in his honor. Sufjan Stevens, a popular recording artist, paid homage to him in his song "Casimir Pulaski Day." Most impressively, in 2009 he became one of only eight people to date to receive honorary U.S. citizenship by an Act of Congress.

8. What have I forgotten to ask about this foreign leader who contributed much of his own money, and, in the end, his life for the American dream and freedom?

Since 2019, Casimir Pulaski has received worldwide attention. His newfound fames comes not for his military accomplishments, but rather as the result of the tests that the Smithsonian Institution did on what most experts believe were his remains. During its investigation, it discovered that Pulaski had a skeletal anatomy that contained many female attributes. Medical experts have surmised that Pulaski likely had a genetic condition known as congenital adrenal hyperplasia, Pulaski was neither male nor female. Rather, they would classify him as intersex. If the skeletal remains are indeed those of Pulaski as seems likely, it is ironic that an American general long remembered for a heroic death might be known best to future generations for his genetic anomaly.

WILLIAM SMALLWOOD

By Charles Wilson Peale, 1781–1782
Independence Hall, Philadelphia (INDE 14148)
Image, public domain

TWENTY-ONE

William Smallwood

"MARYLAND PATRIOT" ★ (1732–1792)

---

1. William Smallwood was a lesser-known figure in the American Revolution, yet gave much of himself. Where was he born, and what do we know about his early years?

We know the birth year for William Smallwood: 1732. But apparently his parents never recorded the specific date of his birth, so like his fellow general Francis Marion, we can only approximate his age. Genealogists believe that Smallwood's earliest American ancestor had moved from Middlewich, England to the colony of Maryland in the 1660s, and settled in Charles County. The Maryland State Archives website states that a location in that county known as "Smallwood's Retreat" was probably his place of birth. It therefore seems unlikely that historians will ever know the exact details of his birth date or place.

2. Is it true that his parents actually sent him to Eton in England for his education. If so, what did he study there, and how did it impact him?

Once settled in Charles County, William Smallwood's first American ancestor became a tobacco planter. His business thrived, allowing him to leave a substantial inheritance for Bayne, his eldest son and William's father. In addition, Bayne's wife brought an ample dowry with her when the two married. Using this wealth, Smallwood's parents were able to send William to a preparatory school in England, and then enroll him at Eton (then, as now, one of the most famous schools in the world).

This institution of higher learning has had a plethora of famous students pass through its doors since the 1440s, including many who became world leaders. One of them, the Duke of Wellington, paid the school the ultimate compliment when he said that "the Battle of Waterloo was won on the playing fields of Eton." But in the case of William Smallwood, historians know very little about his years at Eton, outside of the fact that he matriculated there.

3. At the Battle of White Plains (October 28, 1776), apparently he was wounded twice. What was his contributions there?

Much as we have gaps in our knowledge about William Smallwood's early life, historians have also proved unable to learn with any sense of certainty much about his first foray into the military realm. We know that he served as an officer during the French and Indian War (1754–1763), but we have no specific details of his service. Because most volunteers in the colony of Maryland helped to secure the western part of Maryland around Fort Frederick during that conflict, it is likely that Smallwood participated in that endeavor. Smallwood's first provable association with the military came on January 14, 1776, when Maryland gave him command as a colonel of its 1st Maryland Regiment of volunteers for service in the Continental Army.

By the time his force marched northward to join forces with George Washington, the British had evacuated Boston (March 17, 1776). Correctly anticipating the next British move, Washington concentrated his force of over 20,000 men in the area in and around New York City. Undaunted by this show of strength, in August 1776, the British attacked the American defensive line at the Battle of Long Island, and inflicted a potentially crushing defeat on the Patriots.

Recognizing that he needed to buy time for his retreating force to reform, Washington ordered Smallwood's regiment to seal the breach.

Although it suffered heavy losses, the Marylanders succeeded in slowing the British advance long enough for Washington to safely evacuate the remnants of his army. Unfortunately, the British pursued him to Manhattan, and forced him to retreat northward. In October of 1776, the British caught up with Washington's retreating force, and attacked him at White Plains. In the ensuing battle, the right and center wings of the American Army held firm, but the left wing collapsed when assaulted by a force of Hessians.

This exposed the American center wing, which included Smallwood's regiment, to enemy fire from two directions. Rather than panic, however, these American soldiers kept their structural integrity as they retreated. They abandoned the battlefield, but would live to fight another day. Throughout the battle, Smallwood displayed exemplary skill and courage. Even two wounds that the British inflicted on him did not prevent him from exercising effective control of his regiment. Congress then rewarded his valor by giving him an appointment as a brigadier general in the Continental Army.

4. He apparently served under Washington at the Battle of Germantown (October 4, 1777), and served honorably there. What occurred and transpired?

In the fall of 1777, the British decided that capturing Philadelphia might finally allow them to subdue the Americans. Accordingly, a British army under the command of General William Howe advanced on the nation's capital in September. Examining the terrain between the advancing British and the city, General Washington determined that the banks of Brandywine Creek offered him an excellent defensive position. Placing his men with great care, he awaited a British attack.

Much as he had done at Long Island, however, General Howe maneuvered a force around Washington's defensive position, threatening to rout the Americans. Along with the other units comprising the American right wing, Smallwood's regiment attempted to shift to meet the British flanking movement, but soon found itself compelled to retreat. Only the arrival of a division commanded by General Nathaniel Greene prevented the total annihilation of the American Army. After evacuating Philadelphia in the aftermath of the Battle of Brandywine, in October of 1777, George Washington attempted to make good his loss by attacking the British at Germantown northwest of Philadelphia.

In this battle, a brigade commanded by Smallwood had the greatest success of any American unit and inflicted heavy losses on the British right wing. However, because the other American units did not reach their objectives, Smallwood had to order his men to retreat.

5. Interestingly, he served not just the American Revolution, but his home state of Maryland. What did he contribute there?

After his participation in the unsuccessful campaign to defend Philadelphia, Smallwood next saw action in the Southern theater of the war. Ordered to accompany General Horatio Gates in his effort to regain control South Carolina after the fall of Charleston in 1780, Smallwood and his men participated in the American debacle at the Battle of Camden (August 16, 1780). Although clearly a disaster for the Americans, Camden could have been much worse had it not been for a determined stand that Smallwood's men made as everything collapsed around them. Because his actions had allowed at least a portion of the American force to avoid death or capture, Congress promoted Smallwood to major general.

After the Battle of Camden, he commanded a militia force for a short period of time, but soon went home to Maryland, where the end of the war found him in 1783. His fellow citizens elected him to Congress in 1785, but they also chose him as their governor. Smallwood declined to serve in Congress, but accepted the position as governor. He served three terms of one year each, and then held a seat in the Maryland Senate. Smallwood died in 1792, and years later the State of Maryland made his final resting place a state park.

6. What else should we know about his endeavors and successes?

William Smallwood fought in some very sanguinary battles during the American Revolution with units under his command that often found themselves in literally the thick of the fighting. He obviously fought well, as he received two promotions based on his battlefield efforts. Some readers might point out that every battle he played a major role in resulted in an American defeat, and might therefore feel that Smallwood does not deserve inclusion in any list of great American generals. Such a conclusion would be wrong and a failure to understand a salient feature of the American Revolution.

To succeed, the Americans did not need to triumph consistently on the field of battle: they merely needed to survive. Nathaniel Greene eloquently phrased this principle by saying "we fight, get beat, rise, and fight again." By staving off complete disaster at White Plains and Camden, William Smallwood played just as important a role in the American Revolution as those generals who won battles.

FRIEDRICH WILHELM VON STEUBEN

By Charles Wilson Peale, 1782–1784
Independence Hall, Philadelphia (INDE 14164)
Image, public domain

TWENTY-TWO

Friedrich Wilhelm von Steuben

DRILLMASTER OF THE CONTINENTAL ARMY ★ (1730–1794)

---

1. We know that Frenchman Marquis de Lafayette (1757–1834), serving as a major general and Colonel Thaddeus Kosciusko (1746–1817) of Poland both came to America to assist in fight for freedom and independence, but Friedrich Wilhelm August Heinrich Ferdinand Steuben (born Friedrich Wilhelm Ludolf Gerhard Augustin von Steuben) is one of those unsung heroes of the American Revolution who also served George Washington in many capacities. What do we know of his background, as he obviously was not born in the colonies?

Friedrich von Steuben was born on September 17, 1730, in Magdeburg. At the time of von Steuben's birth, Magdeburg lay inside the boundaries of the Kingdom of Prussia. His father held the rank of captain in the Prussian Army. When the War of the Polish Succession began in 1733, von Steuben's father offered his services to Empress Anna of Russia. Von Steuben accompanied his father to Poland, returning with him when his father left the Russian Army in 1740. After von Steuben rejoined the Prussian Army, he received orders to report for duty at Neisse (now a town in Poland and renamed Nysa). Soon, the army transferred him to Breslau (known today as Wroclaw, and also now a town in Poland).

His father arranged for von Steuben to attend school in both of those communities. Soon, von Steuben's father received orders to accompany the Prussian Army as it invaded the Austrian province of Silesia at the beginning of the War of the Austrian Succession (1740–1748). Von Steuben accompanied his father, and according to some historians von Steuben actually saw action at the age of 14 during that conflict. Even if he did not actually see battle, by his teenage years von Steuben clearly had a great deal of familiarity with the military life.

2. What was von Steuben's military experience before the American Revolution?

Friedrich von Steuben grew up in a military family. Perhaps not surprisingly, he decided at the age of 17 to formally enlist in the Prussian Army. Given the rank of second lieutenant, von Steuben distinguished himself during the Seven Years' War (known in the American Colonies as the French and Indian War, 1756–1763). Wounded twice, and also taken as a prisoner of war, he received promotion to first lieutenant and then captain. His valor brought him to the attention of Prussian ruler Frederick the Great, who assigned him to a special course of instruction for promising junior officers and to whom he served as an aide-de-camp. Von Steuben seemed destined for great things in the Prussian Army. However, when the Seven Years' War ended in 1763, Prussia downsized its military. As a consequence, von Steuben found himself in that year without a job.

3. What brought von Steuben to North America and the 13 colonies and what did he do initially in the Revolutionary War?

When his military career ended in Prussia, von Steuben sought employment in the private sector. In 1764, he secured a position as the *Hofmarschall* of a principality in what is now southwestern Germany. A *Hofmarschall* handled the administrative duties in principality, and usually carried with it a title of nobility. Because of that, he started to call himself a baron in 1771. As part of his duties, von Steuben accompanied his ruler, the Prince of Hohenzollern-Hechingen, on his travels through Europe. These trips brought him into contact with Louis St. Germain, the French Minister of War. In 1777, St. Germain introduced von Steuben to Benjamin Franklin, who was in France to solicit French aid for the American cause.

Franklin immediately recognized that a former Prussian Army officer could prove invaluable to the fledgling American Army, and wrote him a letter of recommendation to Congress. Von Steuben then sailed to North America, went to Congress to report himself a volunteer to the American cause. He was accepted and told to report to General Washington. He arrived at Washington's camp at Valley Forge, Pennsylvania in February of 1778. Recognizing his potential, Washington had him appointed Inspector General, with the rank of major general. He would hold that rank until the end of the American Revolution.

4. Apparently Von Steuben brought much to what we might today call standardization of the military. What were his contributions in this regard?

Upon his arrival at Valley Forge, von Steuben noticed that the American Army seemed to have no grasp of the basic elements of hygiene. His first act as Inspectoral General, therefore, involved creating a system for laying out camps in a manner that put latrines as far away from the cooking facilities as possible. For almost 150 years, the American Army followed this plan when constructing camps. People may remember von Steuben for his role in creating a more effective fighting force, but this earlier contribution probably proved more important in the long run for allowing American soldiers to live long enough to actually see combat.

5. Von Steuben's contributions at Valley Forge were many. Can you enlighten us about his great efforts during that time period?

After improving sanitary conditions at Valley Forge encampment, Von Steuben turned his attention to the training that American soldiers received. He immediately recognized that the Americans had great courage, but only a rudimentary knowledge of how to fight effectively as a military force. For that reason, von Steuben began a program to train soldiers based on what he had learned in the Prussian Army. To accomplish this, von Steuben had American officers recommend the best soldiers under their command for training under his personal direction.

One skill that he wanted the American soldiers to master involved the use of the bayonet. This desire stemmed from the fact that the British had used that weapon effectively up to that point in the American Revolution, while the Americans seemed to have no understanding of its value in close combat. When satisfied that they grasped the essential principles, von Steuben then sent them back to their units to train their fellow soldiers. In this manner, von Steuben proved able in a short amount of time to train a large number of American soldiers.

To further ensure that Washington's troops and future generations of soldiers could benefit from his training program, von Steuben had his instructions put into writing. Officially designated "Regulations for the Order and Discipline of the Troops of the United States," this document was approved by Congress in March, 1779 and became commonly known in the army as the "Blue Book." It remained the American Army's drill training manual until the end of the War of 1812.

6. The Battle of Stony Point is one that needs to be recalled, and remembered. Can you give us the historian's perspective on this, and von Steuben's contributions?

In 1779, the British had embarked on a campaign to widen its area of control in southern New York. As part of this operation, they had created a fortified position at Stony Point on the Hudson River north of New York City. George Washington realized that capturing that position would seriously compromise the British campaign, and ordered an assault on Stony Point in July of 1779. General Anthony Wayne, who commanded the assault, ordered his men to launch a bayonet attack at night.

The Americans executed the plan flawlessly, capturing the fortified position with minimal casualties. Von Steuben's work with the American Army on bayonet tactics had thus proved invaluable to the Americans at Stony Point.

7. Von Steuben's path apparently crossed with John André. What was his role in this event?

John André was a major in the British Army during the American Revolution. When American Benedict Arnold contacted the British about changing sides, André became the intermediary between Arnold and the British. As part of Arnold's betrayal, he had offered to allow the British to capture the American fortifications at West Point on the Hudson River. André, in disguise, had visited Arnold at West Point, but while returning to the British lines a detachment of American soldiers had captured him. When George Washington learned of André's capture, he ordered a military tribunal to try André for espionage. In 1780, von Steuben served on the panel of judges, which found André guilty and sentenced him to death. Von Steuben then served as instructor and supply officer for General Nathaniel Greene's army fighting in the southern colonies and he commanded one of the three divisions of the Continental Army at Yorktown.

8. I understand that there was even a stamp named after him in his honor. How else was this hero honored?

Von Steuben's efforts in organizing and training the Continental Army made him instrumental in the American fight for independence.

While Americans have largely forgotten many of the American generals from the American Revolution, von Steuben, who helped demobilize the army at the conclusion of the war and ultimately died on his farm in the Mohawk Valley of New York, has remained a notable figure. For example, Steubenville, Ohio was named for him. Further, many American cities, primarily in the north and east, host a Von Steuben Day Parade. Interestingly, many American film goers have seen a Von Steuben Day Parade without realizing it, as the famous scene in *Ferris Buehler's Day Off* when the titular character takes over a Chicago parade was filmed during that city's annual celebration of von Steuben. For that reason, von Steuben will always be with us!

JAMES MITCHELL VARNUM

By Charles Wilson Peale, 1804
Independence Hall, Philadelphia (INDE 14152)
Image, public domain

TWENTY-THREE

James Mitchell Varnum

RENAISSANCE PATRIOT ★ (1748–1749)

1. Again, we look at the heroic soldiers who served General Washington and our country during the Revolutionary War, there were many such Patriots. One was James Mitchell Varnum, whose rise to general under Washington can only be described as meteoric! Where was he born and where did he go to school?

James Mitchell Varnum was born on December 17, 1748 in Dracut, Massachusetts, a community located in Middlesex County. Varnum was very smart and gained admission into Harvard College. However, he remained for only a short while at Harvard, choosing instead to transfer to the newly-opened college in the colony of Rhode Island and Providence Plantations—better known today as Brown University.

In 1769, he graduated with honors in the first class to graduate from that institution. Ironically in light of subsequent events, in his senior thesis he argued that the British colonies in North America should remain loyal to the crown. Deciding to remain in Rhode Island after graduation, Varnum initially tried his hand at teaching, but soon began to study for the bar examination and a career in law under the tutelage of Oliver Arnold, the Attorney General of Rhode Island. In 1771, he passed his examination, and gained admission to the bar. Choosing the Providence suburb of East Greenwich to make his home, Varnum established a law office there, and had soon built a successful practice.

Thus, by the age of 22, James Varnum had achieved great success. Little did he know that his star would soon rise even higher.

2. What was this individual like personally?

According to accounts that date from after his death, James Varnum had a magnetic personality and a superior intellect. A chronicle of the early settlers of the Ohio Territory, for example, described him as "a dexterous reasoner, and a splendid orator." In a speech given to commemorate the founding of Marietta, Ohio, a former classmate of Varnum said that his speeches could "reconcile mankind to the closest bonds of society." But perhaps the most telling evidence of the high estimation that Varnum enjoyed from his peers comes from the number of positions of authority, both inside and outside the military, with which he was entrusted.

3. What were his contributions during the Revolutionary War?

Because of the deteriorating relations between Great Britain and its North American colonies, in the fall of 1774 Rhode Island chose to create a number of militia companies in case hostilities broke out between the two sides. One such military entity drew its recruits from East Greenwich. Designated the Kentish Guards, the unit included James Varnum. Upon its official organization, the members of the new company chose Varnum as their captain. Interestingly, the roster of the Kentish Guards included a private named Nathaniel Greene, who would later rise to the rank of major general in the Continental Army and become Varnum's superior officer. After learning of the Battles of Lexington and Concord (April 19, 1775), Rhode Island consolidated its militia companies into three regiments, giving command of the first of them to Varnum. He then led his unit to Boston where it became part of the force keeping the British Army bottled up in Boston.

Once the British evacuated Boston on March 17, 1776, the Continental Army transferred its base of operations to New York, where George Washington correctly surmised the British would make their next move. In an attempt to maximize his chances of holding that strategically important city, Washington kept some of his troops in Manhattan while deploying most of his force to Long Island. Varnum's unit, designated the 9th Continental Regiment since the first of the year, occupied a position in the center of Washington's defensive line. Unfortunately, the British flanked the American position there during what became known as the Battle of Long Island. Given orders to retreat, Varnum led his men in good order from the field of battle. Transported to safety in Manhattan on the night of August 29th, Varnum's men participated in a gradual retreat northward after the British made a successful landing in New York City. Washington's army briefly halted the pursuing British at the Battle of Harlem Heights on September 16, 1776, but his force suffered a decisive defeat at the Battle of White Plains the following month.

This setback forced Washington to retreat in a precipitous manner. He managed to avoid the capture of his entire army by crossing the Delaware River to safety in Pennsylvania in December 1776. Attrition by then had reduced Varnum's regiment to less than 200 men, and because of that he received orders to return to Rhode Island to recruit volunteers.

While back in Rhode Island, Rhode Island promoted him to the rank of brigadier general in its colonial militia. Because of this, Varnum missed Washington's victories at Trenton and Princeton. In February of 1777, Varnum received promotion to the rank of brigadier general in the Continental Army. Command of his regiment—which had regained its designation as the 1st Rhode Island Regiment—then passed to Colonel Christopher Greene.

In his new role Varnum supervised the troops who successfully defended Fort Mercer in New Jersey and Fort Mifflin on Mud Island in the middle of the Delaware River in October 1777, receiving an official commendation for his actions. After wintering with Washington's army at Valley Forge, Varnum next saw action at the Battle of Rhode Island (August 29, 1778). There, troops under his command fought well before eventually abandoning the field of battle.

In March of 1779, Varnum resigned his position in the Continental Army, choosing to resume his prewar legal career. Two months later, Rhode Island promoted him to major general of its militia, and Varnum served in that capacity to the end of the war.

4. Following the war, how did Varnum contribute to the founding of our country?

Varnum had actually started to serve his country politically even before the war ended, as he won election to the Second Continental Congress in 1780. At the end of the American Revolution, Varnum became one of the founders of the Society of the Cincinnati, and organization that only officers who had served in the Continental Army could belong to. In 1786 he returned to Congress, and in 1787 he received an appointment as a judge for the Ohio Territory. After establishing residence there in 1788, Varnum helped write the legal code for the territory. Unfortunately, by then Varnum's health had begun to deteriorate, as he started to display symptoms of tuberculosis. Varnum died from this condition in January 1789, and was buried in Ohio.

5. How do we recognize him today?

In many respects, James Varnum falls into the same category as many of his fellow generals who served competently, but not brilliantly, during the American Revolution. Indeed, with the exception of his role in the defense of Forts Mercer and Mifflin, Varnum did not compile the same record as some of his contemporaries. Having said that, Varnum deserves high marks for a contribution that he made far from the fields of battle.

As previously noted, Varnum's Rhode Island regiment had suffered a significant reduction in force during the first two years of American Revolution, and the state had given him the responsibility of finding individuals to make good the losses. Varnum soon identified a group of male adults that Rhode Island could call upon: African-Americans. His suggestion in one sense did not break ground, as the Continental Army had included African-Americans for the first few months of the conflict. In November 1775, however, the army had adopted a policy of excluding African-Americans from military service. Thus, when Varnum recommended adding African-Americans to the ranks of his regiment, he was simply asking for the reinstatement of a previous practice.

But in retrospect, Varnum's suggestion seems quite radical, because the concept of a segregated army had had two years to become ingrained in the thinking of the American military. Observing military protocol, Varnum gave his recommendation to George Washington, his commanding officer. As a slave owner, Washington may have had some qualms about arming African-Americans, but he nonetheless sent the recommendation on to the Rhode Island state legislature. In February 1778, the members of that body passed an act to allow African-Americans to serve in Varnum's regiment. Free African-Americans simply enlisted in the same manner as white volunteers, but enrolling slaves involved an intermediate step. Once a slave enlisted, he received his freedom, and the state legislature then compensated his owner for his loss.

Estimates vary, but most historians believe that between 100 and 140 African-Americans enlisted in 1778. Throughout the rest of the war, many people referred to Varnum's unit as "the Black regiment," and a famous illustration at the time pictured an African-American dressed in that regiment's uniform. Because of his role in allowing African-Americans to take their rightful place beside their white counterparts in fighting for the independence of the new nation, Varnum should be much better known to the American citizenry.

ANTHONY WAYNE

By James Sharples, Sr., 1796
Independence Hall, Philadelphia (INDE 11922)
Image, public domain

TWENTY-FOUR

"Mad Anthony" Wayne

FEROCIOUS FIGHTER ★ (1745–1796)

---

1. One of the most flamboyant figures of the American Revolution, was, without a doubt, "Mad Anthony" Wayne. Let's start at the beginning of his journey: Where was he born and what were his early years like?

Anthony Wayne was born in Chester County, Pennsylvania, on New Year's Day in 1745. His grand-father and father had both done well in business, and this meant that Anthony Wayne would grow up in affluence. This financial wherewithal allowed Wayne to attend a private school while growing up. He then became a student at the College of Philadelphia (the present-day University of Pennsylvania). Wayne did not earn a degree there, however, choosing instead to return to his family's home in Chester County. As part of his schooling, he had acquired training as a land surveyor, and he soon began to ply this trade (as did General George Washington in his younger years).

Having become a well-known figure in his community, Wane decided to enter politics in 1774 when he stood for election to the Pennsylvania legislature, known in colonial times as the Pennsylvania Provincial Assembly, and after 1776 as the Pennsylvania General Assembly. Wayne won, and assumed his seat in that body. Less than a year later, developments in North America would cause him to give up his seat and adopt a different course of action turning from political service to military service.

2. How did Anthony Wayne first get involved in the Revolutionary War, and what were his first exploits?

Although Wayne left no record of his deliberation on the matter, the news of the Battles of Lexington and Concord in Massachusetts on April 19, 1775, must have had a profound effect on him because he immediately began to recruit a militia regiment in Pennsylvania. He completed his recruitment efforts in the fall of 1775, and on December 9, 1775, this unit received the designation of the 4th Pennsylvania Regiment, and Wayne became its colonel. Pennsylvania then offered this regiment to the Second Continental Congress, and that body ordered the unit to report for duty to General George Washington. He and his regiment were ordered north to Canada and he was wounded at the June 8, 1775 Battle of Three Rivers outside of Quebec at the crossing point of the St. Lawrence River.

3. Often people get a *nom de guerre*—a name of war so to speak. Is there anything in the record that traces when exactly he received the "Mad Anthony" sobriquet?

Numerous accounts from the Revolutionary War era contain passages referring to Anthony Wayne and many of them mention the fact that Wayne had a volcanic temper. Moreover, he had an impetuous streak, especially in the heat of battle. But a close examination of the records indicates that Wayne received his nickname from an incident in 1781. When he recruited his regiment, Wayne had found many willing volunteers in Chester County, and one of them had run afoul of the law while at a tavern. When taken into custody, the soldier said that the authorities should release him because of his close relationship with Anthony Wayne.

Taking the soldier's claim with a grain of salt, the authorities contacted Wayne to check on its veracity. Rather than corroborating the story, Wayne instead said that the soldier should receive an appropriate punishment for his transgression. Furthermore, Wayne said that if the soldier misbehaved again he should receive 29 lashes. Informed of this, the soldier looked incredulous and said "Wayne is mad. Mad Anthony Wayne—that's who he is." The nickname seemed appropriate, and would remain with him to the end of his life.

4. Wayne served in the invasion of Canada by Continental forces. What role did he play?

By the time that the Second Continental Congress ordered Wayne to join the northern wing of the American forces, the effort to capture Canada that had begun in 1775 had already failed. Because of this, Wayne found himself aiding an effort to disengage American forces and withdraw them to safety. During the retreat, an American general named William Thompson received intelligence that a British force stood exposed near Trois-Rivieres, Canada. Thompson ordered an assault, counting on surprise and superior numbers to give him a victory.

Unfortunately for him, he had neither surprise nor superior numbers on his side, and he suffered defeat. The only bright spot for the Americans came from the conduct of Anthony Wayne. Commanding 800 men, Wayne fiercely contested his area of the battlefield as long as he possibly could, and then fought a rearguard action that allowed the remaining American soldiers to retreat. For his heroic actions at the Battle of Three Rivers, Congress promoted him to brigadier-general in February 1777.

5. Wayne fought in three battles during the Philadelphia Campaign of July 1777–July 1778. What was significant about these engagements?

In the summer of 1777, British General William Howe initiated a campaign to capture Philadelphia, where the Second Continental Congress held its meetings. George Washington moved his forces to defend the city, anchoring his forces along the banks of Brandywine Creek. Unfortunately, a Loyalist informed Howe of a ford across the creek that Washington had left unguarded. Using this information, on September 11, Howe managed to put a sizeable force on Washington's flank and launch a surprise attack. While many American units broke and ran, troops under Wayne's command held their ground for three hours. This allowed Washington to withdraw his forces from the battlefield and begin a retreat.

At that point, Wayne took his troops off the battlefield, and deployed them in various communities outside Philadelphia. On the night of September 20, a British force attacked one of Wayne's units near the town of Paoli. Achieving total surprise, the British killed many Americans soldiers with bayonets while they slept. This assault angered Wayne, but soon Washington gave him a chance for revenge. Learning that Howe had quartered a portion of his army near Germantown, Washington decided to attack them on October 4. Ordered to lead the assault, Wayne drove his men forward in a determined charge that broke through the British line.

Unfortunately, Wayne's men advanced faster than the rest of the American Army, and this imbalance allowed the British to counter-attack. After a fierce fight, Washington ordered his men to retreat. Once again, he called on Wayne to hold his position long enough for Washington to fall back, and once again Wayne succeeded in that task. Although Wayne did not win any of these three engagements, his courage under fire earned him great respect on both sides.

6. The Battle of Stony Point was another event that seemed to earn him some fame. What transpired there?

In June 1778, the British evacuated Philadelphia and moved north toward New York City. As they did so, Wayne encouraged General Washington to attack. He did so in what became known as the Battle of Monmouth in which Wayne participated. However, it was another and later battle that was a significant boost to the Continental Army and General Wayne.

In 1779, the British attempted to expand the area under their control around New York City. As part of this effort, a British force seized a site known as Stony Point on the Hudson River. General Washington believed that he had to respond to this move, and decided to order an American force to recapture Stony Point. He gave the assignment to Wayne. Remembering the British attack at Paoli, quickly known as the "Paoli Massacre," Wayne chose to use the same strategy against the defenders. Accordingly, on the night of July 16 he led his men on a bayonet charge, and in thirty minutes his force had captured Stony Point and 550 defenders. In honor of this accomplishment, the Second Continental Congress presented him with a medal.

Later in the war, Washington moved Wayne and his troops south, where he participated in the Southern Campaign of the war under the command of General Nathaniel Greene. There, after fighting Native Americans of the Creek nation who allied with the British, Wayne and his Pennsylvanian troops recaptured and occupied Charleston, South Carolina for the Americans.

7. After the war Wayne also served his country in a political sense. What do we know about his service?

Once the Revolutionary War ended, Wayne returned to Chester County, and his fellow citizens elected him to the Pennsylvania legislature. After serving for a year in that body, Wayne decided to move to Georgia. A few years later, delegates meeting in Philadelphia wrote a Constitution for the nation, and sent the document to the states for ratification. The citizens of Georgia asked Wayne to serve in a state assembly that would decide what action to take regarding the document. Wayne urged ratification, a suggestion that Georgia followed.

In 1791, Georgia elected Wayne to serve in the newly-created U.S. House of Representatives, but some members of that body felt that Wayne did not actually live in the district he represented. Congress ordered Georgia to hold a new election, but Wayne decided that he would not try to keep his seat. He left Congress after the special election, and would never hold political office again.

8. What else should we know about this true American hero and leader, to whom we owe so much?

If people remember Anthony Wayne today, they probably do so primarily because of a military campaign that took place a decade after the end of the Revolutionary War. As Americans moved westward after the conclusion of those hostilities, they met determined resistance from the Native Americans living in those areas. Those wishing to settle in those regions turned to the U.S. government for help, and in 1794 President George Washington decided to use the American Army to subdue the Native Americans who had formed a loose confederacy comprised of the Shawnee, Ottawa, Miami, and Delaware nations.

To lead this campaign, Washington selected Anthony Wayne. Wayne rigorously trained his volunteer force (known as Wayne's Legion), and then led them into the Northwest Territory—present-day Ohio. On August 20, 1794, Wayne decisively defeated a Native American force at a location known as Fallen Timbers. This defeat forced the Native Americans to cede control over a vast amount of territory in the Old Northwest (also known as the Northwest Territory), and allowed thousands of settlers to claim land there.

American history textbooks include a discussion of this battle, but few mention Anthony Wayne in connection with the Revolutionary War. Hopefully, these words will help make people aware of his crucial role in that conflict as well.

Afterword

In 1823, three years before his death, Thomas Jefferson, the third president of the United States, wrote a letter to the then-current and fifth president of the United States, James Monroe. Jefferson wrote of the belief and desire that the new American nation would not be a nation of war. He said that the nations of Europe "are nations of eternal war. …On our part, never had a people so favorable a chance of trying the opposite system, of peace and fraternity with mankind and the direction of all our means and faculties to the purpose of improvement instead of destruction."

Unfortunately, the American Revolution would not be the only war in the history of the United States. Other wars with other leaders—generals and admirals—fill the pages of American history. Those leaders and those wars are also part of the nation's history. But without Washington's generals the history would be very different. It is "counterfactual history" and whether such history might have been better or worse is for the reader to study, think, and decide!

Recommended Reading

The American Revolution and its many facets have generated a remarkable array of reading material for general readers and historians. The works listed below offer those desiring to learn more about the war and its American military leaders good entry points for further reading.

Books about the Revolutionary War

Atkinson, Rick. *The British are Coming: The War for America, Lexington to Princeton, 1775–1777.* New York: Henry Holt, 2019.

Bailyn, Bernard. *The Ideological Origins of the American Revolution.* Enlarged edition. Cambridge, MA: Belknap Press, 2017.

Berkin, Carol. *Revolutionary Mothers: Women in the Struggle for American Independence.* New York: Vintage Press, 2006.

Daughan, George C. *Lexington and Concord: The Battle Heard Round the World.* New York: W. W. Norton, 2019.

Ellis, Joseph J. *The Cause: The American Revolution and Its Discontents 1773–1783.* New York: Liveright, 2021.

Ellis, Joseph J. *Revolutionary Summer: The Birth of American Independence.* New York: Vintage Press, 2014.

Ferling, John. *Almost a Miracle: The American Victory in the War of Independence.* New York: Oxford University Press, 2009.

Gilbert, Alan. *Black Patriots and Loyalists.* Chicago: University of Chicago Press, 2012.

Leckie, Robert. *George Washington's War: The Saga of the American Revolution.* New York: HarperCollins Publishers, 1992.

Levin, Jack E. *The Crossing.* New York: Threshold Editions, 2013.

McCullough, David. *1776.* New York: Simon & Schuster, 2005.

Middlekauff, Robert. *The Glorious Cause: The American Revolution, 1763–1789.* Revised edition. New York: Oxford University Press, 2007.

Philbrick, Nathaniel. *Bunker Hill: A City, Siege, Revolution.* New York: Penguin Books, 2014.

Shorto, Russell. *Revolution Song: A Story of American Freedom.* New York: W. W. Norton, 2017.

Tucker, Phillip Thomas. *Kings Mountain: America's Most Forgotten Battle That Changed the Course of the American Revolution.* Brattleboro, VT: Skyhorse Publishing, 2022.

Wood, Gordon. *The American Revolution: A History.* New York: Modern Library, 2003.

Wood, Gordon. *The Radicalism of the American Revolution.* New York: Vintage Press, 1993.

Books about George Washington and His Generals

Billias, George Athan, ed. *George Washington's Generals and Opponents: The Exploits and Leadership.* New York: Da Capo Press, 1994.

Billias, George Athan, *John Glover and His Marblehead Mariners.* New York: Henry Holt & Co., 1960.

Bragg, C. L. *Crescent Moon over Carolina: William Moultrie & American Liberty.* Columbia, SC: University of South Carolina Press, 2013.

Carbone, Gerald M. *Nathaniel Greene: A Biography of the American Revolution.* New York: St. Martin's Press, 2010.

Chernow, Ron. *Washington: A Life.* New York: Penguin, 2010.

Cole, Ryan. *Light-Horse Harry Lee: The Rise and Fall of a Revolutionary Hero – The Tragic Life of Robert E. Lee's Father.* Washington, D.C.: Regnery Publishing, 2019.

Duncan, Mike. *Hero of Two Worlds: The Marquis de Lafayette in the Age of Revolution.* New York: PublicAffairs, 2021.

Ellis, Joseph J. *His Excellency, George Washington.* New York: Alfred A. Knopf, 2004.

Gaines, James, *For Liberty and Glory: Washington, Lafayette, and Their Revolutions.* New York: W. W. Norton, 2007.

Golway, Terry. *Washington's General: Nathaniel Greene and the Triumph of the American Revolution.* New York: Holt, 2005.

Hazelgrove, William. *Henry Knox's Noble Train: The Story of a Boston Bookseller's Heroic Expedition That Saved the American Revolution.* New York: Prometheus, 2020.

Higginbotham, Don. *Daniel Morgan: Revolutionary Rifleman.* Chapel Hill, NC: University of North Carolina Press, 1979.

Lee, John K. *George Clinton: Master Builder of the Empire State.* Syracuse, NY: Syracuse University Press, 2010.

Lee, Russell David. *The American Revolution in the Southern Colonies.* Jefferson, NC: MacFarland, 2006.

Lockhart, Paul. *The Drillmaster of Valley Forge: The Baron de Steuben and the Making of the American Army.* New York: Harper Perennial, 2010.

Marga, Christopher P. "Soldiers…Bred to the Sea": Marblehead, Massachusetts, and the Origins and Progress of the American Revolution." *The New England Quarterly,* 77:4 (Dec. 2004), 531–62.

Mattern, David B. *Benjamin Lincoln and the American Revolution.* Columbia, SC: University of South Carolina Press, 1998.

Mintz, Max M. *The Generals of Saratoga: John Burgoyne and Horatio Gates.* New Haven: Yale University Press, 1992.

Nelson, Paul D. *The Life of William Alexander, Lord Stirling.* Tuscaloosa: University of Alabama Press, 1987.

Nelson, Paul D. *Anthony Wayne: Soldier of the Early Republic.* Bloomington, IN: Indiana University Press, 1985.

Nester, William R. *George Rogers Clark: "I Glory in War."* Norman, OK: University of Oklahoma Press, 2012.

O'Donnell, Patrick K. *The Indispensables.* New York: Atlantic Monthly Press, 2021.

Oller, John. *The Swamp Fox: How Francis Marion Saved the American Revolution.* Lebanon, IN: De Capo Press, 2018.

Palmer, Dave Richard. *George Washington and Benedict Arnold: A Tale of Two Patriots.* Washington, D.C.: Regnery Publishing, 2006.

Philbrick, Nathaniel. *Valiant Ambition: George Washington, Benedict Arnold, and the Fate of the American Revolution.* New York. Vantage, 2016

Piecuch, Jim and John Beakes. *"Light Horse Harry" Lee in the War for Independence.* Baltimore, MD: Nautical & Aviation Publishing Company of America, 2013.

Puls, Mark. *Henry Knox: Visionary General of the American Revolution.* New York: Palgrave Macmillan, 2008.

Royster, Charles. *Light-Horse Harry Lee and the Legacy of the American Revolution.* Cambridge, UK: Cambridge University Press, 1982.

Simms, William Gilmore. *The Life of Francis Marion* rev, ed., Greenwood: WI, 2019.

Storozynski, Alex. *The Peasant Prince: Thaddeus Kosciuszko and the Age of Revolution.* New York: St. Martin's Press, 2010.

Taaffe, Stephen R. *Washington's Revolutionary War Generals* by Stephen R. Taaffe. Campaigns and Commanders Series, Volume 68. Norman: University of Oklahoma Press, 2019.

Tucker, Thomas Phillip. *Saving Washington's Army: The Brilliant Last Stand of General John Glover at the Battle of Pell's Point, New York, October 18, 1776.* New York: Skyhorse Publishing, 2022.

Unger, Harlow Giles. *Lafayette.* New York: Wiley, 2003.

Valentine, Alan C. *Lord Stirling.* New York: Oxford University Press, 1969.

Zambone, Albert Louis. *Daniel Morgan: A Revolutionary Life*. Yardley, PA: Westholme Publishing, 2018.

About the Authors

Donald Elder, Ph.D., is professor of history and chair of the History Department at Eastern New Mexico University in Portales, New Mexico. His undergraduate degree in history is from the University of Northern Iowa. He received the M.A. and Ph.D. from the University of California, San Diego. He began his history career as a historian of the American Space Program and currently concentrates his research efforts on the American Civil War. He is the author and editor of numerous books and articles.

Michael F. Shaughnessy, Ph.D., is a professor of educational studies at Eastern New Mexico University in Portales, New Mexico. He holds a bachelor's degree from Mercy College in Dobbs Ferry, New York, and two master's degrees and a doctorate from the University of Nebraska in Lincoln, Nebraska. He has edited, co-edited and written approximately 30 books and more than 500 articles, research pieces, and book reviews.